Astronomers' Universe

For further volumes:
http://www.springer.com/series/6960

Ben Moore

Elephants in Space

The Past, Present and Future of Life and the Universe

Ben Moore
Centre for Theoretical Astrophysics and
Cosmology
Institute for Computational Science
University of Zurich
Zurich, Switzerland

ISSN 1614-659X ISSN 2197-6651 (electronic)
ISBN 978-3-319-05671-5 ISBN 978-3-319-05672-2 (eBook)
DOI 10.1007/978-3-319-05672-2
Springer Cham Heidelberg New York Dordrecht London

Library of Congress Control Number: 2014940552

Translation from the German language edition: *Elefanten im All – Unser Platz im Universum* by Ben Moore,

Printed on acid-free paper

Springer is part of Springer Science+Business Media (www.springer.com)

In the memory of my father
For Mariana and Joe

Foreword

The Romance of the Universe

It is hard to think of a more romantic story than the story of our universe itself.

I realise that many will find this statement incredulous. After all, where is the sex and violence?

On a superficial level one could respond to this question quite simply: The history of life on Earth provides more examples of quirky sex than most people could imagine on their own, and nothing humans have ever invented can compare with the violence and explosive power of a supernova, the most brilliant cosmic fireworks in the sky.

But when I used the word romance here, I was referring more to the romance of science itself. It is perhaps not properly appreciated that what drives most scientists to spend the better part of their intellectual lives pursuing sometimes seemingly esoteric questions is not to the desire to save humanity, or even the natural human capacity self-aggrandisement. It is instead a love affair with the universe: the desire to understand something that no one before has understood; to push the frontiers forward so that new questions can be asked about phenomena that may have been previously invisible; to understand how it is that we came to where we are; and to comprehend realistically what the future might bring.

It is this love affair that Ben Moore captures in this book about the history of the universe. Ben has spent the better part of his career exploring the incredibly complex gravitational interplay of objects to form the structures that we know and love as we look out with our telescopes, from galaxies and clusters of galaxies, to stars and planets. When the resources didn't exist, he built his own supercomputer to help carry out the complicated numerical algorithms necessary to explore questions that are simply too complex to handle analytically with standard mathematical formulations. What has characterised his theoretical efforts, in my experience as a collaborator, observer, and friend, is a sense of adventure and fun that permeates much of what he does in his life, and that permeates this enjoyable volume. I encourage you to jump in headfirst. Don't be afraid. You will be delighted and enlightened.

Tempe, AZ
2014

Lawrence M. Krauss

Preface

My father was a forester and spent much of his life immersed in nature from sunrise to sunset. During the school holidays I would join him as he worked tirelessly and thought constantly. He instilled in me his love of the natural world and his wonder in its origins. What are all those ants thinking? How do trees defy gravity and suck water from their roots into such high canopies? Why does the Earth spin and orbit the Sun like some kind of mechanised clock? What is the nature of light? Can black holes really exist? Are we alone in this vast universe or is it filled with life?

This book is written in the memory of my father, who encouraged me to go to university to find some of the answers to the questions we discussed. It is for anyone who wonders about the world around them and who is interested in its origins. It is for my children to read, to inspire them in the same way as my father inspired me. It is something I would like to leave behind in this short but incredible life.

I want to tell the story of the history and future of life and the universe at a level that anybody who is interested in it can understand and enjoy. I will start at the beginning and end at the end, describing our place in time and space, how we got here and where we are going. I will take you on a journey from the beginning of time to the end of the universe to uncover our origins and glimpse at our destiny.

It is a journey of science fiction proportions but based on 3,000 years of scientific findings. I will explain how we acquired this knowledge, beginning with the ancient Greeks who pioneered the art of scientific investigation. This takes us on a remarkable path of discovery from the origin of atoms to dark matter and dark energy, from ants and elephants to space travel and life beyond our solar system.

The following is a true story based on actual events that took place. It's a good time to tell it, since in the last decade astrophysicists have collected new observational data that allow us to understand in detail the history of our universe and how all of its contents emerged. I have spent the last 25 years carrying out research on many of these topics and I would like to explain to you all of these developments using simple and friendly non-technical language. You may encounter some concepts that are difficult to understand—don't worry, some are rather tricky to explain. There is still a lot to learn and in Chap. 10 I will be honest about what we don't understand.

I have always been curious as to the lack of evidence for intelligent life elsewhere in the cosmos. There seems to be nothing special about our star, the Sun. There are billions of similar stars within the Milky Way, itself one of countless other galaxies. The goal of one of our most exciting research projects is to understand the origin of planetary systems and to determine whether habitable planets are abundant or rare. Our supercomputer simulations predicted that 'Earth-like' planets around stars similar to the Sun should indeed be very common. In the past few years very exciting observations have emerged from the Kepler space satellite that is discovering vast numbers of distant new worlds. I wonder how many of these have atmospheres and climates suitable for the emergence of life? How many of those worlds have civilisations more advanced than our own and have already begun to explore the galaxy?

Elephants are amongst the most intelligent creatures on Earth. They love, they mourn, they are social and fun, and their brains have a larger computational capacity than does the human brain. Without our Moon, life on Earth might have evolved rather differently, and elephants not at all; I will explain why later. I have recently given many popular and research talks entitled 'The frequency of elephants in the galaxy' which tie together all of these topics. Our latest research suggests that our galaxy alone hosts at least a billion planetary systems suitable for the development of intelligent life. I suspect that there are many other planets with wonderful thinking creatures like elephants; we just haven't looked in the right way or in the right places.

Our Earth has made over four and a half billion orbits around the Sun. If we can avoid self-extinction and collisions with giant asteroids, we can survive literally for billions of years as a species. Indeed, whilst humans have developed the frightening capability to cause mass extinction events, the same technology will be necessary to protect our planet from the inevitable impact of a rogue asteroid. Such random catastrophic events are rather like the random walks that our lives seem to follow—a journey during which apparently haphazard occurrences can dramatically alter our path. Just spend a few moments in contemplation after reading these words, and the future of your life could be very different.

Our brief time of consciousness is very special, but ultimately, we face the realisation that nothing lasts forever. The most recent observations from our great observatories, together with the theoretical work of my cosmologist colleagues, enable me to discuss the future of the universe. As time ticks on, and the stars begin to fade away and stop shining, can life continue for eternity or will the future be eternally dark? And what is the purpose of it all anyway?!

Zürich, Switzerland Ben Moore

Contents

Chapter 1
A Brief History of Human Knowledge and Discovery

B. Moore, *Elephants in Space*, Astronomers' Universe,
DOI 10.1007/978-3-319-05672-2_1, © Springer International Publishing Switzerland 2014

The sky appeared as if it had been painted a dazzling crimson as the Sun sank below the forested skyline. A line of posts stretched into the distance following the curves of the ancient glaciated valley. The post driver lay on the floor, a primitive, hand-welded iron cylinder used to manually pound fence poles into the ground. Two pairs of bare hands gripped the cold handles and, in synchronised action, lifted the weight high above the wooden post then brought it smashing down. The ground shook and the iron resonated like a church bell, until the next blow struck three seconds later. With each stroke, the splintered wooden post sank a little further into the Earth. As my father and I worked, our minds were elsewhere. We were contemplating the strange duality of light, the bizarre fact that the behaviour of photons could only be understood if they existed simultaneously as both waves and particles. We had reached this question after discussing the reason why molecules in the air preferentially scattered blue light, resulting in our evening's spectacular red sky.

We live in a unique era in the history of humankind. We believe that we have accurately determined our place in the universe, in both time and space. We have realised our insignificance as we gaze out on 100 billion galaxies similar to our own, each containing billions of stars like our Sun. We have measured the size of our universe and determined its age. We know when our Sun started to shine and when it will die. We have found that our universe is expanding; the space between the galaxies is growing, and its rate of expansion is rapidly increasing. We have measured the tiny irregularities in the very early universe, like the ripples on the surface of a lake—the initial conditions from which all the stars and galaxies that we see have slowly emerged over cosmic time. It is a remarkable accomplishment of the human species that we have a good understanding of the history of our universe all the way back to one millionth of a second after it came into existence. At that moment, our entire visible universe and everything in it, all of its matter and energy, was squeezed into a region the size of a football stadium. That sounds amazing, and it really is.

Over the past four hundred years there has been an explosion of knowledge and understanding. It should have begun two thousand years ago. We have come a long way, yet I believe that we could have gone much further. Many important and fundamental aspects of our universe remain a complete mystery. If I could have the answer to one and only one question, it would be, What happened in the first millionth of a second after the big bang? From this knowledge I would surely gain insight into why and how our universe came to be. Our understanding is lacking, not because the question is too difficult to answer but because our current knowledge is limited. It might not even be possible to answer the question with any certainty since we cannot test our theories with observations or experiments prior to that extraordinary event. But we should certainly continue our quest to understand more about our physical universe, our origins and our destiny.

What cosmology has taught us over the past hundred years, and in particular over the last decade, has allowed us for the first time in history to predict how our universe will evolve in the future. Philosophers have long struggled to explain the meaning and purpose of life. I would like to consider this topic within a broad scientific context that encompasses some of our collective global knowledge, from physics and cosmology to neuroscience and biology. I want to take you on a remarkable journey of discovery that allows us to understand our place in the universe, how the universe evolved to yield conditions hospitable for the development of life, and how those conditions are becoming ever harsher and more inimical to life, until ultimately... well, at this point I refer you to the latter part of this book, which deals with our future.

The long-term survival of the human race rests on a systematic quest for knowledge and discovery especially during the next few hundred years. I can only dream of the findings that will emerge over the next millennium, let alone over longer reaches of time. It is, in principle, possible for our species to survive and to continue living on Earth for about another billion years. Probably no longer than this, though, for reasons that will become clear after I explain how the Sun evolves with time and eventually dies.

What is left to be learned to enable us ultimately to answer the question, Why? We can describe matter and energy, gravity, space and time with equations, yet at the deepest level, our understanding of all of these fundamental components of our universe is rather limited. I would like to illustrate one of the pinnacles of human achievement with a one-sentence summary of the theory of general relativity by Albert Einstein: "Time and space and gravitation have no separate existence from matter." What an insight, a unified description of gravity as a geometric property of four-dimensional space-time—the brain child of perhaps the greatest scientist in history. But there are still some fundamental things left to understand. For example, the description of how particles behave in time and space is given by the theory of quantum mechanics, about which American physicist Richard Feynman said, equally succinctly, "I think I can safely say that nobody understands quantum mechanics." Quantum mechanics and general relativity are separate theories that describe the very small and the very large. During the early universe, everything was squeezed into a tiny volume, and under these conditions quantum mechanics and gravity were important on the same scale. Thus, if we are to understand further back in time towards the instant the universe appeared, we will need a "grand unified theory" that combines all the fundamental forces of nature, including a unification of quantum mechanics and gravitation.

Before going any further into the workings of our universe as we understand it at this moment, let me digress into the story of human history that shows the path by which we accumulated that knowledge. The timeline of evolution and comparison to other species will reveal our place in the scheme of things on Earth. I will also lay the foundations for perceiving what our civilisation and our minds might truly be capable of. Let us take a look at how we got this far. How did we manage to figure out that the world could be comprehended? At what point did some people begin to discard mysticism, gods and spirits and to explain natural phenomena as a consequence of cause and effect?

Let us divide human history into several eras based on the rate of acquisition of knowledge, perhaps the most important trait and commodity of an advanced civilisation. The first phase of human history lasted for several million years in which our ancestors spent most of their time as hunter-gatherers. The second great era began around 30000 BC, when art and music appeared and modern *Homo sapiens* started to explore the world both physically and mentally. Agriculture and settlements proliferated from about 10000 BC. The first burst of scientific discovery followed, beginning in earnest with the ancient Greek philosophers and scientists in 1000 BC and culminating in a million scrolls of knowledge shelved in the great libraries of Alexandria. This first remarkable period of intellectual innovation ended with the beginning of the Roman Empire around 100 BC, at which point a long period of scientific listlessness set in. Human attention turned instead to conquest and battle, and law and order. Religious idealism flourished. Interest in the physical world was crushed and suppressed, only to be awakened in another burst of brilliance in the seventeenth century, which constituted the second age of scientific discovery.

Twilight: 3 Million–30000 BC

The first evidence that our distant ape-like ancestors walked upright on two feet can be seen in bipedal footprints preserved in volcanic ash in Kenya. These steps were taken several million years ago by the Australopithecines. The hominid species had emerged and was similar in physical form and appearance to modern humans. It slowly evolved and branched into distinct but closely related subclasses over time, including *Homo neanderthalensis* and, finally, *Homo sapiens*. One of the first Neanderthal skulls to be discovered was found in the Neanderthal valley near Dusseldorf, just three years before Charles Darwin, in 1859, published *The Origin of Species by Means of Natural Selection* (alternately titled *Preservation of Favoured Races in the Struggle for Life*).

In this famous study Darwin wrote "Although much remains obscure, and will long remain obscure, I can entertain no doubt…, that the view which most naturalists entertain, and which I formally entertained—namely, that each species has been independently created—is erroneous." Darwin goes on, species by species, to discuss his ideas and thoughts. His book reveals that his ideas stemmed from the complex interrelationships that he observed between species. He writes about pigeons and domestic animals, of plants and flowers. The idea of evolution by natural selection was profound. Curiously, Darwin avoids discussing the origin of humans in this book but treats the topic at length in *The Descent of Man* (1870), a two-volume work in which he proposes that man may have evolved from apes.

At that time, the lineage of missing links—that is, the fossil record—had not been discovered. In fact, Darwin's ideas were based on speculation, not evidence, and were far from a complete or proven theory. Much of his evidence for man's descent from apes was complete nonsense and based on the differences between Europeans and the "savages" of Africa and Australasia who "could not count above four". In *The Descent of Man* he wrote: "Judging from the hideous ornaments, and the equally hideous music admired by most savages, it might be urged that their aesthetic faculty was not so highly developed as in certain animals, for instance, as in birds." He would, he noted, "rather be descended from monkeys than the barbarian savages of Africa and Australasia." However, his instinct was correct:

the final words of his book read: "Man still bears in his bodily frame the indelible stamp of his lowly origin."

Around 200,000 years ago, *Homo sapiens* appeared in Africa and began to diverge rapidly from the rest of the animal world. All of the branches that led to our species became extinct, except for modern humans, who are left as the only living species in the *Homo* genus of bipedal primates in Hominidae—the great ape family. The human DNA record shows that we actually diverged from Neanderthals almost half a million years ago. Nevertheless, *Homo neanderthalensis* and *Homo sapiens* did interbreed shortly before Neanderthals disappeared (just before the end of the last period of great glaciations, between the Palaeolithic and Mesolithic eras), and a small percentage of the genetic record of Europeans and Asians was contributed by Neanderthal relatives. Interestingly, Neanderthals were taller, stronger and had larger brains than early *Homo sapiens*, yet they developed very little advanced behaviour beyond tool making and were not a communal species.

The first renaissance in human intelligence began around 30000 BC, at which time something happened to make humans start to use their brains and begin to think. The rate at which they acquired skills and knowledge exceeded that of any other life form on Earth. Art, music, social order, agriculture, writing, science, philosophy and engineering all developed at a breathtaking pace, culminating with the work of the Greek philosophers, who questioned the origins of life, the atomic structure of matter and even the nature of stars and civilisations beyond their own. The foundations of modern humans had truly begun.

Why at that time? Why *Homo sapiens*? Throughout its 4.5-billion-year history, Earth has experienced numerous cataclysmic events, such as giant asteroid impacts and dramatic climate changes. But there have also been long periods of climate stability lasting millions of years. Many species have been present on Earth for much longer periods than *Homo sapiens*, yet they did not develop our mental skills. The other extant great apes, our genetically close relatives, show many traits of human intelligence, such as the use of tools and symbols, mourning and ritual, empathy and self-awareness. But we are alone in our ability to understand the natural world and to exploit its natural resources to benefit our existence and to propagate our species. What was it that led to the rapid creative development of our species? The answer is as yet unknown. Evolution appears to halt once a species finds a stable niche. Ants provide a telling example. Millions of years ago, they developed skills similar to those of early humans, such as farming, social order and communication. Yet they have barely evolved since.

Evolution together with natural selection through inheritance of favourable traits has given rise to an extremely diverse assortment of life on the planet that is interdependent in many different ways. However, the average lifetime of a mammalian species is only about one million years. Entire genetic lines disappear when they become unable to adapt to changing conditions or lose out in the competition for resources with another fitter and better-adapted species. Of all the species that have ever existed on Earth, it is estimated that over 99 percent are extinct. Indeed, evolution follows a haphazard and non-optimal path. Each step in the chain is based on small differences from the previous step while at the same time incorporating the legacy design of parents. Eventually, a species can become so different that it bears little physical resemblance to its distant ancestors. But its architecture still contains remnants of its distant past.

The evolutionary biologist Richard Dawkins describes the laryngeal nerve of the giraffe as a striking example to show that animals still exhibit remarkable evidence that they (and we) descended from fish. This particular nerve connects the larynx to the brain and controls the breathing and swallowing of all animals as well as the sounds that they make. In a giraffe the nerve stretches all the way down its long neck, around its heart and back up to its throat. This is a four metre journey that could be made in a few centimetres by simply travelling directly from the larynx to the brain. Similarly, in humans this nerve passes from the brain all the way down to the chest and back up to the throat. The reason for this unnecessarily long path is because the nerve originally evolved in fish. It took the shortest journey from their brains to their gills, and their gills lie near their hearts. As species evolved, necks became longer and so did the nerve, but it still took the long route around the heart. An intelligent designer would at some point have stopped and rethought this scheme to optimise it. There would have been no need to design a nerve ten times longer than it needed to be. The laryngeal nerve shows how evolution must deal with the body plans left behind by previous generations.

Evolution thus does not necessarily represent the optimum path for the survival and advancement of life. Species evolve in a fashion similar to new releases of a computer program, which has to maintain and provide support for older versions. Computer operating systems often become clumsy and slow because they are overloaded with outdated support for old components that are rarely used and have been superseded. Life evolves step by slow step, generation by generation, striving to improve its design for survival and reproduction while at the same time maintaining support for body parts that may have been essential in the past. It cannot just start afresh. Most new branches of life die out; others may survive for millions of years. Sometimes, legacy products are just partitioned off, such as the appendix, which has no apparent use nowadays and is thought to be a relic of a digestive tract. Other body parts can be put to alternative use, such as a penguin's wings, which are no longer used for flight but instead have slowly evolved to aid underwater swimming. An advantage of artificial life over biological life is that it could optimally design and construct its next-generation state. It could start afresh and fill each new machine with a completely original operating system specifically designed for its brand new host.

So why have humans developed so far and not dinosaurs, dolphins, ants or elephants? Did humans just get lucky?

From what we know of evolution, it is not surprising that we share much of our genetic composition with other animals; over 90 percent of our DNA is identical to that of a chimpanzee. However, the small difference in DNA compared to the large difference in our mental capabilities appears to be extraordinary. As our distant ancestors moved from the trees to the plains and developed a diverse and easy-to-acquire diet, for millennia humans evolved as nomadic hunter-gatherers. They began to use their brains beyond the day-to-day tasks of survival, eating and reproduction. They developed the capacity to communicate more efficiently, to plan in advance, to see the benefits of social organisation, to co-operate and to solve problems.

Some information for the basic behaviour of animals is thought to be passed down genetically since brains seem to emerge with pre-encoded algorithms that keep hearts beating and tell lungs to breathe. However, most of our skills and actions are based on knowledge that we must learn, beginning with our time in the womb and for many years after we are born. Our advancement owes itself to our brains being able not only to store and process our learned knowledge but also to think creatively and come up with new ideas and insights. Those rare inventions and ideas that change society and the course of human history often originate with a solitary individual. If we look back over the past 3,000 years, for which we have a good documented record of human achievement, we find that it is the steady accumulation of knowledge combined with the exceptional insight of one person that leads to sudden important advances in culture or scientific understanding. A genius can change the course of history much more than the actions of the average masses. A list of unique people would include household names from among the ancient Greeks—Archimedes, Aristotle, Democritus and Euclid—continuing with great intellects from the seventeenth century onwards, such as René Descartes, Galileo Galilei, Isaac Newton and Einstein. Such people are extremely unusual.

However, the creativity of such minds is based on the accumulated knowledge that is the product of generations of previous scholars. It would have been impossible for Newton to have deduced general relativity and derive the same equations as Einstein. Likewise, Einstein could hardly have conceived all the mathematics and principles of physics that he needed to formulate his theories. He made a massive leap when he linked gravity, space and time, but he could not have done it without the groundwork of earlier scientists. Prodigies, too, are constrained by their surroundings and the store of human knowledge available.

Progress also requires a critical mass of population to drive it. A few hundred or even a few thousand members of a species living together are less likely to produce and nurture a genius. Indeed, the number of cultural traits in chimpanzee communities correlates with the number of females in the population: Female chimpanzees use tools more frequently than the males, they spend more time with their young and they transfer novel cultural traits from other communities. Since new discoveries are based on older knowledge, major advances could not occur until humans ended their hunter-gatherer lifestyle and began to establish permanent settlements. Once those settlements became large enough to connect and communicate with each other, ideas and knowledge could be shared and built on. Progress is a runaway process, but it required the invention of a few basic tools. One of those was controlled fire.

The first undisputed evidence for the systematic use of fire dates back about 200,000 years, although it may be as old as a million years. Fire pits and fragments of burnt chards of bones and clay heated to over 600 degrees centigrade have been found in Africa, Europe, Asia and China, showing that the use of controlled fire was already widespread among hominids one hundred thousand years ago. Genetic change can take place on a timescale of just a few thousand years. We know this from looking at how humans have adapted to recent dietary modifications, such as the consumption of animal milk and the consequent development of lactose tolerance, and how specific genes have developed to cope with a new diet of grain resulting from agriculture. Our brain is only 2 percent of our body weight, but it uses over 20 percent of our energy resources. It has been argued that certain cooked foods were easier to digest and gave a richer source of calories, allowing the brain to develop further and increase in size.

Perhaps it was a combination of nurture and nature that led to our unique progress within the animal world. Humans developed the ability to grow food and store it through hard times, thus preventing starvation and death. They were able to pass information from parent to child, down through generations, allowing the continuation and accumulation of knowledge even through periods of harsh environmental changes or natural disasters. Climate change, giant asteroids and other extreme events wiped out entire species in a very short timescale. If a species were not completely decimated, an existing population could still be damaged to such an extent that it would revert back to instinctual knowledge alone. The difference with *Homo sapiens* was that they could maintain their societies, at least in a basic form, as well as pass on existing knowledge through periods of hardship.

The development and complexity of language points to its origin in Africa some 50,000 years ago. Most languages are formed from about 40 distinct sounds, or phonemes. However, recent research by New Zealand biologist Quentin Atkinson showed that the number of phonemes in different languages decreases the further away from South-West Africa the language is spoken, consistent with the idea that the complexity of a language correlates with its age. English has 45 phonemes, while Hawaiian (at the far end of the human migration route out of Africa) has only 13. The San bushmen of Namibia have lived as hunter-gatherers for as long as records exist, and they use over 100 distinct sounds in their language, including a variety of click noises. They are the focal point of the highest phoneme count, implying that Namibia and surrounding regions could be the place from where *Homo sapiens* migrated outward. The world map of language complexity follows the migration pattern of humans and is remarkably similar to the world map of human genetic diversity, which is also highest in South-West Africa. The DNA of the Bushmen shows a divergence from the rest of humanity that split off some 70,000 years ago.

Perhaps it was this early development of language and social organisation that enabled our species to survive through global natural disasters, such as the enormous volcanic eruption that took place 70,000 years ago at Lake Toba in Sumatra, Indonesia. The explosion had a magnitude of M8 or "mega-colossal", which is at the top of the volcanic explosive index. It created a huge caldera larger than the area of London, and ejected 3,000 cubic kilometres of molten magma into the atmosphere, covering Asia in 15 centimetres of ash. In some places in India the ash was up to six metres deep. The resulting volcanic winter lowered the Earth's mean surface temperature by several degrees, resulting in a 1,000-year cold period followed by the last great ice age. Evidence for these colder temperatures comes from studying gases trapped in ancient ice, which provide a record of the average temperature at the time the ice was frozen.

Genetic data suggest that the entire human population was reduced to a size of about 10,000 breeding pairs between 50,000 and 100,000 years ago. The volcanic explosion at Lake Toba may have been the cause. Humans almost became extinct but some survived in Africa. It is to this small group of survivors that we owe our existence today. By the end of the Late Palaeolithic period, around 10000 BC, the entire population of the Earth is estimated to have stabilised at a total of several million people.

Awakening: 30000–146 BC

It took a long time to develop the conditions that gave our predecessors time for thought and creativity beyond the need for survival. The transition from apes to *Homo sapiens* took over one hundred thousand generations before we began to walk out of Africa. This era of awakening began when the ability to see, appreciate and express the natural beauty in the world was made and preserved in art, and when humans began to enjoy music. In 2008, archaeologists from the University of Tübingen discovered a very early example of human figurative art in a cave called Hohle Fels near Baden-Württemberg. The Venus of Hohle Fels is a six centimetre high sculpture of a woman carved into the ivory of a mammoth tusk over 30,000 years ago. In the same cave, the researchers found a bone flute, also dating back to the same period, which makes it one of the oldest-known musical instruments. These ancient items are evidence that early humans had time for thought, expression, creativity and enjoyment. The first representational art (and what beautiful art it is) started to appear at a similar time in Spain and France. Particularly impressive is the Palaeolithic art in the caves of the Dordogne in France. Created deep underground, painted by the light of burning candles, the scenes depict majestic animals in flight. Of over 900 animals drawn, from birds to bison, 600 can be clearly identified. It is not known why the animals were painted and why so deep underground. My brother has spent most of his life living in the Dordogne. As he showed me the beautiful wide rivers flowing through limestone cliffs lined with natural caves and oak forests, it became clear why some of our very distant ancestors settled there.

The last Ice Age ended around 10000 BC. Thereafter, the first signs of organised agriculture appeared in numerous different locations, from such far-flung places as Syria and China, and spread rapidly over the following few thousand years. The ability to cultivate plants and herd animals may have become possible thanks to the warmer climate following the Ice Age. Mesolithic megaliths appeared at Stonehenge and Nabta Playa in Egypt between 5000 and 3000 BC. These extraordinary monuments had close connections with rituals such as burial and also with the stars and the seasons. The people who constructed them appreciated the cycles of nature; these giant calendars were a testament to their organisational and engineering capabilities. The monuments were built at around the time that the wheel was invented and even before the first written words appeared, scribed on Sumerian clay tablets dating from 3000 BC.

During the latter part of our evolution as hunter-gatherers, the global population stabilised with the birth rate approximately equal to the death rate. However, as technology and science progressed, the population grew rapidly. At the time we synchronised our calendars relative to events in Bethlehem, the population is estimated to have been between 200 and 400 million people. It took until 1800 AD for the Earth's population to grow to one billion people, but the next billion was reached by 1930, then three billion by 1959, four billion by 1974 and five billion by 1987. The pattern is clear... The current global birth rate is about 20 babies per 1,000 people per year, whilst the global death rate is just 10 per 1,000 people per year. That is about four births and two deaths every single second of every day, which does not balance. Our population is still increasing rapidly and may already have exceeded the maximum that our planet can comfortably sustain.

The Greek islands and the shores of the Aegean Sea provided an ideal environment for abundant food, prosperity and technological advancement. These circumstances gave its citizens time for intellectual and creative thought. Most ancient Greeks believed that the Earth was carried around the Sun by a mythical god, that the gods were responsible for all natural phenomena from earthquakes to the weather, and indeed for most aspects of the world and their lives. However, a few individuals began to question the workings of the natural world—they tried to explain things through cause and effect, through observations and experiments. One of the first recognised philosophers was Thales of Miletus in the sixth century BC. He came from an ancient Greek Ionian city that is now part of Turkey. Thales' versatility was remarkable; he was a mathematician and the creator of Thales' theorem which describes how right-angled triangles can be constructed from three points on a circle when two of the points lie on the intersection of a line through its centre. Thales was also a businessman and a politician. One story recounts a year in which Thales bought all the olive presses in the town after predicting a good harvest for the year—the first known example of options trading!

Thales thought about the fundamental constituents of matter and believed, as did the Babylonians, that the world began as water. Whereas the Babylonians held that their god, Marduk, created the dry land, Thales explained the drying up of the water by drawing comparisons to natural processes that he observed around him. Even though many of his ideas were incorrect, his method of thinking and his original approach formed the basis of scientific investigation today. We do not accept irrational solutions which may not be tested and proven to explain what we observe. We are not afraid to question orthodoxy, and we accept a given theory only if it continues to fit our observations.

Anaximander, a student of Thales, continued this free-thinking philosophy and believed that physical forces, rather than supernatural means, created order in the cosmos. He wrote the first-known document about the universe and the origins of life. He conceived of a mechanical model of the world and carried out the earliest recorded scientific experiment: Using the moving shadow cast by a vertical stick, he found that he could determine the length of the year and thus the occurrence of the seasons. These first philosophers and scientists made a brave step forward from a supernatural world, based on imaginary ideas, into the real world, based on support from observations and scientific experiment. Their theories and ideas were obviously incomplete, and they longed to understand more. They did not have the wealth of written knowledge that we have today. They had to start at the very beginning. It must have taken even more courage to abandon a mythical explanation for a physical explanation. It is so much easier just to say, "It is the will of the gods," than to think through the complex chain of events that explain why things really happen. Even faced with the formidable task of understanding the origin of life, these thinkers were willing to stretch their imaginations. Anaximander studied the profusion of living things and their interrelationships and concluded that life started in water and mud and then colonised dry land. He wrote that human beings must have evolved from simpler forms. This insight had to wait 24 centuries before the theory of evolution via natural selection was proposed by the British naturalists Alfred Russel Wallace and Charles Darwin.

The Ionians were engineers and inventors; their legacy is tools and techniques that remain the basis of modern technology. They experimented with air, which could not be seen, and determined that it must be made of something so finely divided that it was invisible. These ideas were carried forward by Democritus (460–370 BC). He was the first to understand that all things are made of tiny particles so small that you cannot see them. He called them atoms. He believed that a large number of other worlds existed, wandering in space, and that they form and also die continuously, some with life, others dry and barren. He was the first to realise that the night sky is filled with stars like our Sun. In his mind, man and the Earth were just a microcosmos. Democritus correctly envisaged the place of humans in the universe. He believed that life was to be enjoyed and understood. What an insightful and beautiful vision into the purpose of life, a vision that I share. He believed that nothing happens at random, that everything has a material cause. He was a devoted seeker of knowledge, and is claimed to have said, "I would rather understand one cause than be King of Persia." He believed that the religions of his time were evil and that neither souls nor immortal gods existed. Luckily, the ancient Greeks were tolerant of these "alternative" explanations of the natural world.

Throughout history, creative individuals have been able to find solutions to problems that at first appeared to be very difficult to solve. For example, how would you prove that the Earth is round? How would you measure its size? There is a much simpler way than photographing it from space or travelling all the way around it and returning to where you started. In 240 BC, a brilliant Greek named Eratosthenes did it with a small stick, a calendar and some simple geometry. Not only did he show that the Earth was spherical, he also measured its size and calculated the answer accurate to within 10 percent. Eratosthenes realised that the Sun was directly overhead in one place each year at the same time, so that if he went to a different place at the same time the following year, a vertical stick would cast a shadow. Thanks to another Greek named Pythagoras who invented some elegant trigonometry, the length of the shadow, together with the distance between the two places, could be used to measure the size of the Earth.

There is a long list of ancient Greeks who contributed to the quest for knowledge and understanding. Of course, the ideas that abounded during these early explorative centuries were characterized not only by brilliance but also by strangeness, including mystical notions such as those advocated by Pythagoras. He recognised that the Earth was a sphere but thought that a strange mathematical harmony was the basis of all nature. Whereas Anaximander and Democritus believed that nature could be explained by experimentation and observation, Pythagoras believed that the laws of nature could be understood by pure thought alone. The followers of Pythagoras were mathematicians and mystics who believed that all things could be derived from numbers and shapes. Aristotle, a student of Plato, wrote about motion and weight and even carried out experiments on falling objects, just like those of Galileo two thousand years later. However, Aristotle's results were inaccurate, and his conclusions and ideas on these topics were incorrect. Unfortunately, his metaphysical thinking had a strong influence on the western world until it was finally replaced by Newtonian physics.

Coinciding with the advances in mathematics, geometry and the sciences, the Greeks invented many things that today we take for granted in our everyday lives. The long list includes tunnels and aqueducts, winches, cranes and water mills, locks and gears, lighthouses, plumbing, showers, central heating, urban planning, odometers and alarm clocks. The Greeks' impressive knowledge can be seen in a single surviving object, the incredible Antikythera mechanism, which was recovered in 1900 from a shipwreck that occurred sometime between the first and second centuries BC. It is an accurate mechanical computer that keeps track of a host of astronomical events; nothing came even close to its sophistication until the late sixteenth century AD. Its degree of complexity and precision has even been compared to a nineteenth-century Swiss clock. The mechanism was corroded and incomplete, but 30 gears have been identified, and it is thought that it may have contained many more.

Using X-ray tomography, a method of imaging devices without damaging them, scientists have been able to reconstruct the functions of the Antikythera mechanism and read part of the 2,000-character manual engraved on it. Upon entering the date, the mechanism calculated the position of the Sun and Moon. Its calendar divided the year into 365.25 days and compensated for the extra quarter of a day in the solar year by using a leap year every four years and then adding an extra day to February. This is the foundation of our calendars today and predates the Julian and Gregorian calendars. What is left of the inscribed user manual refers to the other known planets, and the device may have also indicated their motions. It contained an almanac for the rising and setting times of certain stars. It could show the 19-year Metonic cycle which is a common multiple of 235 synodic months (the period of the Moon's rotation with respect to the Sun over which the phases of the Moon appear to change). Another dial showed the 18-year Saris cycle, the time between occurrences of a particular solar eclipse. Just the Moon's mechanism alone used an ingenious gearing system to show its position very accurately, since the Greeks knew that the speed of the Moon varies over the lunar month (due to its slightly elliptical orbit about the Earth). And in addition to all of that, it kept track of the dates of the main ancient Greek Olympic games, which alternated on three and four year timescales. This remarkable device was constructed more than 2,200 years ago!

Ideas continued to flow from 1000 to 100 BC. If the pace of technological advancement had continued, who knows but that the Industrial Revolution would have started at least 1,500 years ago. Google could have appeared at least a thousand years earlier, and by today we might already have begun interstellar travel, found cures for cancer and even finally understood the opposite sex. That would have all been really great. But no, thanks to the Romans and the spread of religion, however, the increase in global knowledge not only came to a standstill, it was burnt and destroyed and a great deal of rational thought was suppressed for centuries. Most of the knowledge stored in the great libraries of Alexandria was lost forever, pillaged and set to flames during the reign of Julius Caesar. The Roman leaders cared only for obedience, war and domination. At the same time some people had the smart idea that others could be controlled through fear, ignorance and indoctrination. The knowledge and ideas of the first pioneering scientists and philosophers confronted the very foundations and teachings that underpinned the ideas of a creator, of a god, of the church. This is my personal and, some may say, rather harsh assessment of these scientific dark ages. But it was an inevitable consequence of human greed and a desire for wealth and control.

Intellectual Darkness: 146 BC–1600 AD

Ancient Greece was conquered and assimilated into the beginnings of the Roman Empire after the battle of Corinth in 146 BC, during which over 25,000 soldiers from the Roman republic fought and defeated 14,000 Greeks. The domination of neighbouring countries took place at a rapid pace and continued to the point where the Empire occupied a large part of the known world by 100 AD. When the Romans conquered a new country, they cleverly embraced its religions and incorporated them rather than imposing the ancient Roman religious idealism. They applied their resources to maintaining borders rather than investing in science. Nevertheless, they did make significant advances in engineering and mechanisation, greatly improving on the ideas developed by the ancient Greeks. Unfortunately, the network of roads and waterways that they constructed to support the distant reaches of the Empire also promoted communication and thus the proliferation of religion.

The Western Roman Empire collapsed in 476 AD when Flavius Odoacer, a Germanic general in Italy, disposed of the Roman emperor Augustulus. The Eastern Roman, or Byzantine, Empire ended in 1453 AD with the death of Constantine XI and the capture of Constantinople by the Ottoman Turks. Following the decline of the Roman Empire in the West, there was a long era of intellectual darkness, and for a sustained period of time there was little progress of any scientific or mathematical value.

The good times had by the Romans in the East began a slow decline at a time which coincided with various natural disasters, such as the Justinian bubonic plague in 541 AD. Thousands of people died each day in Constantinople, and almost half of the population of the Mediterranean countries was wiped out. Emperor Justinian had a particularly unlucky reign, which suffered many other natural disasters such as floods and earthquakes. The church spread the message that this was god's judgment. The explanation was accepted by many, and the church grew considerably in size and wealth. People felt the need to please this powerful god who was frowning on them, and they wanted to avoid punishment and hell. The attitude indicated by the Christian writers during this period parallels the common fourteenth-century interpretation of the Black Death—the idea that it, too, was caused by the wrath of god.

Imagine what it must have been like 10,000 years ago given the limitations that early people had to face, an ever-present quest for survival yet some spare time to think and ponder their origins. Would you have thought about how you got here? What the twinkling stars were? How night and day arose? What causes the seasons? Of course you would have, but would you have come to the conclusion that something beyond your control was responsible? Perhaps you might even have conjured up the concept of a powerful force controlling nature and ruling your destiny. Without the knowledge that humans possess today, you may well have dreamed up some "almighty being" conveniently responsible for everything.

From 100000 BC ancient humans began burying their dead, a custom often cited as the first evidence of religion. Perhaps it was simply a sign of respect that also provided a place where one could go to mourn and remember loved ones. Atheists nowadays follow the same practices. The first evidence of a place of worship dates from around 10000 BC at the fascinating excavation site of Göbekli Tepe in Turkey. This is around the same time that the foundations of Stonehenge in England were constructed. These sites were used as gathering places to appreciate and celebrate the cycles of nature. The pyramids and the first religious texts appeared in Egypt and Greece in 2400 BC. We find words about Osiris, the god of the underworld or afterlife. The ancient Hindu Vedic religion was also founded during this period. From 2000 BC the Minoans of the Greek island of Crete exclusively worshipped female gods, the bringers of life. Indeed, many religions seemed to begin with matriarchal worship but evolved into patriarchal systems. Most religions have a divine creator and placed humans and the Earth at the centre of the universe. Being at the centre of everything made humans feel important. On the other hand, some religions, such as Buddhism, define "a way of life" and do not attempt to answer the question of how the universe came to be. In fact, the Buddha said, "Conjecture about [the origin, etc. of] the world is an unconjecturable that is not to be conjectured about, that would bring madness and vexation to anyone who conjectured about it."

At least some of the ancient Greeks had a fair idea that humans were not at the centre of anything, and that the Earth orbited around the Sun, the so-called heliocentric view of the cosmos. In fact, it was Aristarchus (c. 270 BC) who first measured the relative sizes of the Earth, Moon and Sun, as well as the distances between them and the order in which the known planets moved around the Sun. However, he did not know the distances to the other planets. That had to wait until the invention of the telescope in the seventeenth century, even though the first glass-making manual was written in 650 BC! Aristarchus also knew that the stars were very far away since there was no detectable "parallax": the apparent movement of the positions of the stars over the course of an entire year as the Earth orbits the Sun. His writings on the heliocentric cosmos have unfortunately not survived, but some information was passed down to modern historians and scientists through his contemporaries such as Archimedes. The Greeks made huge advances in determining the place of Earth in the cosmos. These ideas lay dormant, however, until the sixteenth and seventeenth centuries, when scientists spoke out against the teachings of the church.

Standing outside and observing the sky makes it seem logical that the Earth is stationary and that everything is moving around us. It was also easy to mislead people into thinking that the Earth was at the centre of the universe since they were pulled towards the centre of the Earth. During the Dark and Middle Ages, the idea that the Earth was moving and spinning around the Sun was labelled nonsensical since people and objects would fall off! This is a logical but incorrect argument. If the Earth were spinning about 18 times faster, we would indeed start to float, although at that speed the Earth itself would break into pieces. People do not like what they do not understand and naturally seek a solution, any solution, even one that has no real basis. So it may be understandable that people wanted to believe that there was some overall creator and guiding force behind the world and the universe. Many people like this idea of being at the focus of everything—a very special place because, well, just because people like to feel special and to think that the universe is the way it is just for our benefit. Similar ideas are entertained by some scientists today, under the name of the Anthropic Principle, which reminds me of ideas that proliferated during these scientific dark ages.

The church played a notable role in suppressing the idea of a heliocentric world throughout the Middle Ages, as epitomised by the story of Giordano Bruno, a fifteenth-century astronomer and philosopher. Like Democritus, Bruno imagined that the stars in the night sky were identical to the Sun and that they, too, hosted their own planetary systems. To him, the universe was infinite to reflect the infinity of god. His seven year trial by the church ended in the year 1600 with Bruno being burned at the stake for heresy. The surviving summary documents of his trial clearly indicate the confrontation and suppression of science by the church.

The ideas formulated by Aristarchus were finally recognised and advocated strongly by the Polish mathematician and astronomer Copernicus in 1566 in his major work *On the Revolutions of the Celestial Spheres*, which started people discussing astronomy and science again. A fundamental principle of cosmology today is the Copernican principle, which states that the Earth is not in a central, special place in the universe. Aristarchus should receive credit for it. Copernicus was aware of his ideas, referencing Aristarchus in an early draft of his book, although the reference was removed from his final version. The ideas of the ancient Greeks, that the stars were distant "suns" took until the seventeenth century to become accepted again.

In 1616 church officials were asked to decide whether or not to accept the notion that the Earth actually moves around the more massive Sun. Accordingly, Galileo was summoned to Rome to examine the pros and cons of heliocentric astronomy. The church was willing to accept that the universe was heliocentric as a calculating device, but not as literal truth. Pope Urban VIII initially encouraged Galileo to publish the arguments for and against heliocentric astronomy, but quickly regretted this decision when Galileo became too vocal about the issue. Galileo was arrested for ostensibly making fun of the Pope in 1633. He was put on trial for his ideas, found guilty and sentenced to house arrest. It was not until 1992 that the Catholic Church revoked its condemnation of Galileo and apologised for his mistreatment.

The mathematical and scientific revolution began in earnest in Northern Europe in the seventeenth century, when scientists began to displace the religious ideals of the time, despite continued conflict with the church. This was the age in which the universal mechanistic laws of nature were established, two thousand years after Anaximander and Democritus.

Enlightenment: Seventeenth Century Onwards

In Neo-Platonic Christianity, the planets were carried by angels and spirits based on Aristotle's erroneous notions that whatever moves must be moved and that constant motion requires a constant force. In Aristotelian physics, weight was an inherent property of an object, and its natural state was to be at rest. These properties of matter and motion were based on naive observation, but perhaps one that most of us would also make based on our everyday experience. Two clear-thinking geniuses, Galileo and Newton, completely changed our understanding.

It all began with Galileo, who was performing controlled reproducible experiments on the motion of bodies to infer that they always fell to Earth, accelerating at a rate equal to 9.8 metres per second each second *independent of their mass*. This new "universal law of nature" ruled out Aristotle's notions of weight. Galileo also discovered the fundamental concept of "inertia", the tendency of matter to preserve its state, whether it be at rest or in motion. Descartes generalised this concept to assert that the natural state of motion of an undisturbed body in the universe is not at rest, or in circular motion, but moving in a straight line at constant speed. This was the foundation of physics! Without this insight, Newton's famous laws of motion could not follow.

Not long before Newton, algebra was expressed in words. The idea of writing symbols to represent unknown quantities and irrational numbers was not developed until Descartes and Pierre de Fermat in the early seventeenth century. Previously, doing mathematics with the Roman numeral scheme was inconvenient, to say the least. For example, $65 - 16$ is $\mathrm{LXV} - \mathrm{XVI} = \mathrm{XLIX}$. It was not until around the fourteenth century that it was replaced by the Hindu-Arabic system—the one we still use today—which employs ten digits and includes zero. In the 50 years before Newton, mathematics was sufficiently developed to allow him to invent differential calculus and to devise his famous three laws of motion and the law of gravitation. Newton's unified theory of mechanics was based on experiments with falling bodies carried out by Galileo and the empirical laws of planetary motion discovered by Johannes Kepler. The fundamental idea of universal mechanistic laws of nature was first advocated by Descartes and finally realised by Newton.

The principle of inertia, or Newton's first law, states that if something is moving with nothing touching or disturbing it, it will keep on moving at the same speed in a straight line forever. Newton's second law describes how one must exert a force to change the speed or direction of motion of an object. Newton's third law is the action-reaction principle, of cause and effect—forces are not caused by properties of empty space. It turns out that Newton's first and second laws are a consequence of the symmetrical nature of space. These symmetries are not like the reflection symmetry of a mirror. They come from the fact that motion is independent of direction and orientation. The symmetry of time means that the laws of physics do not depend on time. This gives rise to the principle of conservation of energy and Newton's third law.

Newton's first law seems strange to us since whenever we set something in motion it never travels in a straight line and eventually comes to a standstill. The reason is that the force of gravity pulls the object towards the centre of the Earth, causing its trajectory to curve, and friction against the air molecules that are constantly bouncing off its surface will cause it to slow down and stop. The total energy is still conserved; the energy in the motion of the object has been transferred to the motions of the molecules of air, which become hotter. The nature of inertia and Newton's second law of motion are neither intuitive nor obvious. If they were, persistent thinkers such as Plato and Aristotle might have discovered the correct mathematical laws of nature long before.

New ideas often follow from the interpretation of observations and experiments. Tycho Brahe, a Danish nobleman, spent most of his life observing the stars and making notes on the motions of the planets. He also observed the appearance of a new star—a supernova, as well as a comet, which indicated that the heavens were not invariant, an idea very much disliked by the church. Brahe was a careful and committed observer of the night sky and documented the accurate positions of the planets over many years. He was rather secretive and did not even allow his assistant, Johannes Kepler, access to his notes. But after Brahe died, Kepler found and carefully studied his data for Mars. After much effort, he showed that Mars lies on a precisely repeating but non-circular elliptical orbit about the Sun.

Even though Kepler sent his results to Galileo, who failed to make the connection, there was no unification of their separate results and ideas. . .until Newton. Newton went on to show that this trajectory of Mars was exactly as would be expected were there a "force of gravity" acting across the space between the Sun and Mars. He proved that the strength of this force decreases as one over the square of the distance between them. If two objects are moved twice as far apart, the force between them becomes four times (2^2) weaker. When they are three times further apart, the force becomes nine times weaker.

Newton thought that the same mysterious invisible force must act between the Earth and Moon to keep it in its motion about the Earth. Otherwise, it would travel in a straight line away from us. The force of gravity from Earth must cause the Moon to fall back to Earth, but at just the right amount to compensate for the distance it moves away from the Earth in the same instant of time. And it does. How can it be constantly falling without getting any closer? This is easier to see if you imagine dropping a ball and measuring the time it takes to hit the ground. Now throw the ball horizontally above the ground. It will still fall to the ground in the same time no matter how fast you can throw it, since the gravitational force from the Earth only affects the ball in the vertical direction. However, if you could propel the ball fast enough, at about 8 kilometres per second, the Earth's surface will curve away from the ball at the same rate at which the ball falls back to Earth. In the absence of friction from the air, the ball would continue to travel around and around the Earth, constantly falling but moving forwards quickly enough that it stays the same height above the Earth's surface.

Using Newton's inverse square law of gravity, we can calculate how far the Moon must fall in one second knowing that the acceleration must be weaker by a factor which is the square of the ratio of the Earth's radius to the distance between the Earth and the Moon. The acceleration on the Earth's surface is such that in one second an object falls about 4.9 metres. If we calculate the numbers, we find that the Moon falls about one millimetre towards the Earth every second, during which time the Moon has moved forward one kilometre. Because the Moon orbits the Earth about 385,000 kilometres away, this is exactly the distance it needs to "fall" to Earth each second so that it stays the same distance away from us. When Newton first did this calculation to test his force law, the numbers did not quite work out precisely enough for him. Disappointed, he delayed publication of the *Principia* for several years, at which time a more accurate measurement of the size of the Earth was made and his predictions turned out to be correct.

The success of Newton's theories heralded the beginning of modern predictive science. "If I have seen a little further it is by standing on the shoulders of giants." Newton may have made this famous quip, from a letter dated 1676 to his shorter rival Robert Hooke, as a joke. However, there is no doubt that Newton stands alongside Einstein in his intellectual contributions to science. Newton asked the question, What if all bodies attract each other with an invisible force? What a question. And what an answer he found! His work marked the beginning of the present era of discovery and gave us the first predictive framework within which we could begin to answer those long-standing questions about our origins.

Several hundred years later, Einstein refined Newton's theory of gravity, and founded a more general theory of gravity that incorporated the curvature of space and time by matter and energy. If any theorem or universal law fails in any way in its description of nature, then it can only be incorrect or incomplete. Newton assumed that the gravitational force was instantaneous. Einstein developed a more general and accurate theory under the assumption that gravity, like light, travels at a finite speed, the limiting speed of light beyond which nothing can travel faster. In 1905 he published one of his many classic papers, "Does the Inertia of a Body Depend on Its Energy Content?" He derived his famous equation, $E = mc^2$, which showed the equivalence of energy and mass. They are the same thing, related by a simple proportionality constant (the square of the speed of light). Stating the law of conservation of energy is identical to saying that mass is always conserved.

Since Newton, progress in our understanding of the natural world has increased dramatically and constantly over the past 400 years, from the macroscopic scale of the entire visible universe, to the microscopic inner world of atoms. Now and again, a single individual, in making bold leaps, can change the way we think about the universe. But science is built on generations of thought and discovery, of wrong paths and mistakes, of luck and serendipity. Today there are many more scientists than ever before. We argue and cross-check our results, and we are happy to find flaws in the theories of our colleagues. A member of the general public wandering into a scientific conference session might be surprised by the rough-and-ready exchange of ideas and critical judgement, though it is usually constructive. Indeed, in an age of "big science", collaboration is an integral part of the research enterprise.

Despite our progress, however, there is much that we do not know. Many of the open questions we pose today are the same as those asked over the previous millennia, including the most fundamental: Why is the universe here? Why does it have the properties it has? How did it begin? We are still at a stage in our evolution and capabilities where science has not yet achieved a deep enough understanding to provide the answers, but scientists should be encouraged to work towards this goal. Before we can understand the first and most difficult questions, which start with the words *why*, *how* and *what*, we need to gain a deeper understanding of some of the fundamental aspects of our universe, including the nature of space, time and gravity...of dark matter and dark energy. There are other fascinating questions to which we seek answers, which also seem daunting. I will attempt to answer some of them from both a scientific and personal viewpoint in this book: What is our place and purpose in the universe? Are we alone? and Can life exist forever?

Working outside with my father in the midst of nature in the remote Northumbrian and Yorkshire valleys and woodlands was a very special experience. I was very fortunate to have such a man as my companion and early mentor. He was a free thinker who questioned everything around him. He wanted to know how things worked. He was asking the same questions posed by generations of people before him, the same questions that had puzzled Democritus two millennia ago, "How did the universe come into existence? How did all this complexity and order emerge?" As we talked, it seemed that more questions than answers arose, instilling a desire within me to understand my physical surroundings and the origins of the universe, the nature of light and matter. University seemed a good place to aim for, to study the knowledge left behind by all the great scientists, the legacy of our civilisation.

Chapter 2
What We Know and How We Know It

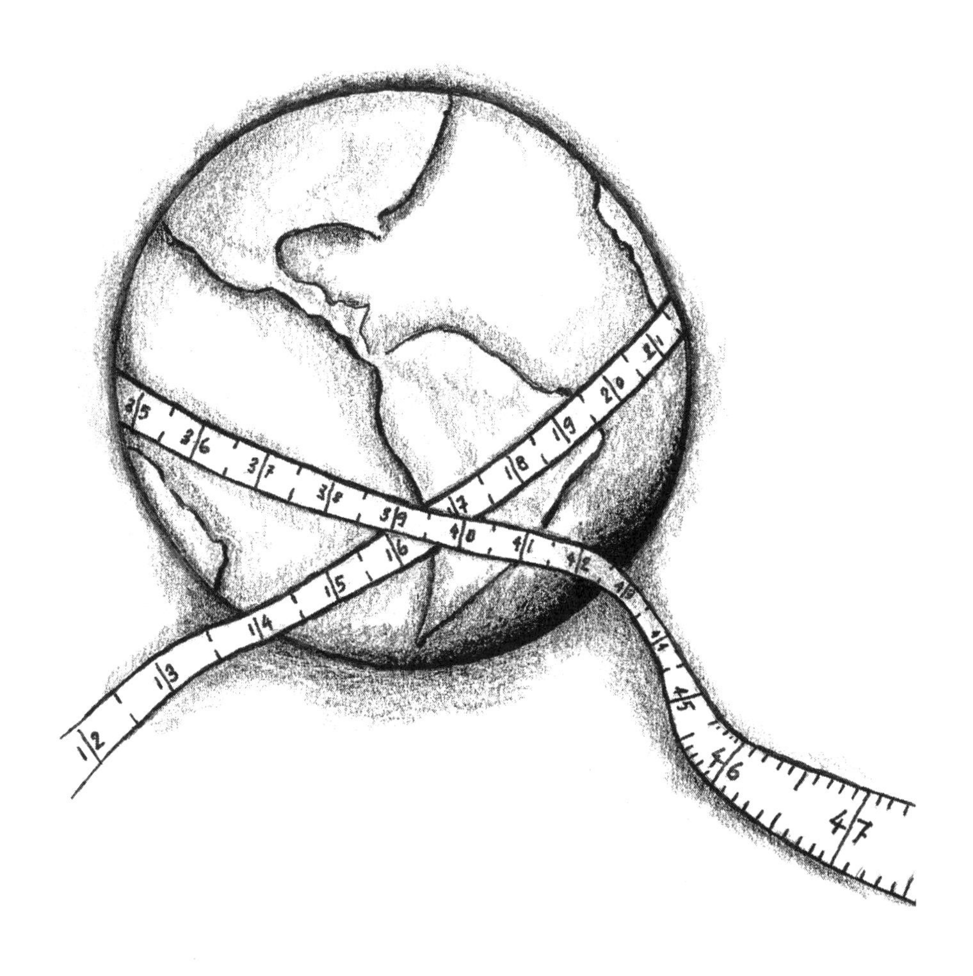

B. Moore, *Elephants in Space*, Astronomers' Universe,
DOI 10.1007/978-3-319-05672-2_2, © Springer International Publishing Switzerland 2014

During my school holidays as a teenager, I made sandwiches, served food and washed dishes in my sister's café in Whitby on the Yorkshire coast while her husband toured Europe with reggae band Jab-Jab. I was saving up enough money to buy a guitar so that I could fulfil an early dream of playing in a rock band. Days were long, starting at 8 am and ending at 8 pm, and nights were spent playing Galaxian and Tron in the brightly lit amusement arcades that still line the waterfront today. The highlight of my early musical ambitions was playing "Paranoid" by Black Sabbath in front of the entire school with a thousand children and parents watching. My best friend, Fletch, sang in a voice that was remarkably close to that of Ozzy Osbourne, and the event was marred only by his ex-girlfriend throwing a custard pie in his face at the end of our act.

The influence of our teachers and mentors cannot be overstated. I was just 15 when my physics teacher, Mr Jones, informed me that my capabilities were abominable. That was after I was engaged in a particularly humorous conversation with Fletch which was unfortunately overheard by Mr Jones. My limitations for a career in science were laid bare in front of the class as I was told to stand in the corner with my back to everyone "like a shaggy dog outside the supermarket". The humiliation was intense, but it instilled in me a desire to prove my teacher wrong. Despite this setback, it seemed to me that it would be much easier to become a scientist than a rock star, although the desire to play music has never left me. My musical ambitions were fulfilled when I was invited to play guitar with the electro-rock band Milk67. It has been a huge amount of fun, culminating in our "Big Bang Truck" at the 2010 Zurich Street Parade. Our lovemobile was 25 metres long and carried a hundred dancing people dressed for the party, a laser harp and a million-volt musical Tesla coil. The back of the lovemobile was a wall of speakers—a hundred thousand watts of sound that shook the buildings as we drove slowly around the lake while we played live to half a million people. The combination of science and music is fun!

The scientific revolution started by the likes of Galileo, Descartes and Newton never slowed down, and it is still continuing. The greatest achievement of our species is the knowledge that we have acquired which enables us to understand our natural world and the cosmos. We have accurately determined the age, size and extent of our universe and have established how it has evolved over most of its history. It is no longer considered a theory that the universe is expanding, or that it is as old and as large as it is. This understanding is widely accepted as fact among the scientific community, and by anyone who familiarises themselves with the large amount of evidence. This all sounds impressive, but how can we be so sure? How do we measure the ages of stars and distant galaxies that we cannot visit? How do we measure the size of the universe across which we are unable to travel? How do we know that the universe is expanding and that it has a finite age, and thus a definite beginning? Are there any viable alternative explanations? In this chapter, I will try to answer these and other questions.

To comprehend our place in the universe we need to appreciate and understand the universe as a whole. We certainly do not know all the answers, and there are many open questions in astrophysics and cosmology, just as there are in other scientific disciplines. For example, we have been trying for many years to make a cosmic inventory of all the matter that exists anywhere to answer the rather fundamental question, What is the universe made of? The stars and galaxies that we can see with our telescopes are not the largest component of matter in the universe—far from it. We know that the atoms of which stars are made, and of which we are made, are only a small fraction of the matter that exists in our universe. Since we know that matter and energy are the same thing, and that both affect the history and future of the universe, we must take all forms of matter and energy into consideration in our cosmic inventory. Two of the biggest problems in understanding the composition of the universe are the unknown nature of dark matter and dark energy. I will get to the first of these "dark" topics later in this chapter. Before we continue, let me describe the convention that scientists use to discuss the enormous range of numbers that they need to quantify the universe.

This table shows the number system in use throughout most of the scientific world. The metric system was introduced at the end of the eighteenth century under the instruction of King Louis XVI of France with just six prefixes, from a thousandth to one thousand. In 1960 the scale was extended from a billionth to a billion. In addition, the International System of Units was defined, known as "Système International d'unités" in French, denoted by "SI".

Power of ten	Number	Common use	Prefix
10^{-18}	0.000 000 000 000 000 001	Quintillionth	a (atto)
10^{-15}	0.000 000 000 000 001	Quadrillionth	f (femto)
10^{-12}	0.000 000 000 001	Trillionth	p (pico)
10^{-9}	0.000 000 001	Billionth	n (nano)
10^{-6}	0.000 001	Millionth	μ (micro)
10^{-3}	0.001	Thousandth	m (milli)
10^{-2}	0.01	Hundredth	c (centi)
10^{-1}	0.1	Tenth	d (deci)
10^{0}	1	One	
10^{1}	10	Ten	da (deca)
10^{2}	100	Hundred	h (hecto)
10^{3}	1,000	Thousand	k (kilo)
10^{6}	1,000,000	Million	M (mega)
10^{9}	1,000,000,000	Billion	G (Giga)
10^{12}	1,000,000,000,000	Trillion	T (Tera)
10^{15}	1,000,000,000,000,000	Quadrillion	P (Peta)
10^{18}	1,000,000,000,000,000,000	Quintillion	E (Exa)

It is convenient to write numbers in powers of 10. For example, one of the most fundamental "constants of nature" is the speed of light, which is 299,792,458 metres per second, which can written (approximately) as 3×10^8 metres per second. This means that in one second light will travel a distance of almost 300,000 kilometres, which is almost the distance from our Earth to the Moon. Scientists often round numbers up or down since often it is just the "order of magnitude" (or general size) that really matters. In one year, light travels 10 trillion kilometres. A trillion is 1,000,000,000,000, which is a lot of zeros to write out and count, so we simply say 10^{12}. The universe is so large that we measure distances in terms of this length: the "light year". It takes light just a fraction of a second to cross the entire Earth and only eight minutes to reach us from the Sun, but it takes over four years to reach us from Alpha Centauri, the next closest star to us. You can already imagine that our own Milky Way galaxy is rather large. The smallest objects we deal with are those of individual elementary particles, such as the electron, which has an effective classical size of 0.000000000000001 metres, or 10^{-15} metres, which is one quadrillionth of a metre. Appreciating numbers so large and so small is difficult, to say the least. We are used to amounts that have a very different and limited scale. We are familiar with seeing a crowd of one hundred thousand people in a stadium, and we can imagine a city of a million people. Once we reach a billion, our imaginations start to fail and we basically lump everything larger as just. . .well, very, very large.

Time

To measure the age of something, we need to define a scale of time. The rotation of our Earth and its motion around the Sun give us the natural timescales of a day and a year. The Babylonians divided the day and night into 24 hours. They subdivided sexagesimally, one sixtieth of an hour, and one sixtieth again, although the first clock to actually display seconds did not appear until the latter half of the sixteenth century. This definition of time, calculated from the motion of the Earth, did not change for several thousand years. It was used as an international standard until 1967. However, subtle variations in the length of the Earth's day mean that this definition of time changes with time. Consequently, defining a second by the rotation of the Earth is not a precise enough or stable standard of measurement. Furthermore, even the best mechanical clocks and watches are still only accurate to a few seconds a day.

The length of our "day" is actually a consequence of an event that took place over four billion years ago during the formation of our solar system. A small planet, about the size of Mars collided with our Earth and because the collision was not quite head on, it set the Earth rapidly spinning on its axis and sent a cloud of debris into orbit. This led to the formation of our Moon, and the Earth has been spinning ever since! Our "24-hour day" is an outcome of action and reaction, and the collisions that took place during the early era of the solar system could easily have been different, resulting in our day being ten times shorter or a hundred times longer.

That's right—an object that is rotating in space will continue rotating forever until something stops it. This is true for an object spinning in space or an object that is kept in orbit by the force of gravity, such as the motion of the Earth around the Sun. It is similar to Isaac Newton's first law of motion, that something moving through empty space will continue moving in a straight line and at the same speed forever until some external force acts on it. A moving or spinning object has an intrinsic energy, called kinetic energy, which is always conserved. You can extract this energy and convert it into heat, which is another form of energy, but in the process the object will slow down and eventually stop. This is how hydroelectric power or wind power works. The reason that the water or the air came to be in motion and have this energy that we can extract is because of the Sun. As it radiates its energy into space and onto Earth, it evaporates water from the oceans which then rises high into the warm atmosphere. The air circulates from the equator to the poles, and the water eventually condenses and falls as rain or snow onto land and mountain tops, a cyclic process from which we could satisfy all of our energy requirements via wind or hydroelectric power.

Our Earth spins perpetually in the near vacuum of empty space. It is a natural clock that never needs winding. If we tried to extract energy from the Earth's rotation, it would result in the planet's spin slowing down and the day becoming longer. In fact this is happening very slowly. Due to the gravitational force of the Moon, which causes our ocean tides, the day is getting longer at a rate of about two milliseconds per century. The friction of the ocean's waters sloshing over the Earth's surface results in some of the rotational energy being transferred into "waste" heat. We can and do use the motion of ocean waves to generate energy with specially made turbines. How much energy could be harnessed from the spin of our planet? The average power consumption of the entire human race is about 20 terrawatts; one light bulb requires about 100 watts of power, so this is equivalent to leaving 200 billion light bulbs constantly lit. The total energy that is stored in the Earth's rotation is immense, enough to provide this much energy for billions of years. But if we extracted all of it, the consequence would be a lengthening of the day until the Earth spins no more.

The fact that the Earth is moving around the Sun at 30 kilometres per second gives the planet kinetic energy of motion that is 10,000 times larger than the energy stored in its rotation. This is a source of energy that is hard to tap into—can you think of a way? I hope not, because if you extracted the energy of the Earth's motion through space, it would slowly spiral into the Sun. The fact that constant motion is well preserved is a good thing. Since it formed, our Earth has made 4.5 billion journeys around our star; the hominid species began walking upright just a few million of those revolutions ago. As it orbits the Sun, the Earth has rotated through night and day over three trillion times. Early *Homo sapiens* lived a short life of only about 9,000 days, whereas today the average human can enjoy about 26,000 days of which I have already lived more than half of my allocation. Life for us humans is short.

There is an ever increasing need to measure time, and other constants such as length and mass, with higher and higher accuracy. We now define time using an atomic clock, and more specifically, one second is the interval between 9,192,631,770 wavelengths of photons (packets of light energy) that are emitted by a particular isotope of the element caesium. Microwave photons are shone onto metallic caesium atoms, causing them to vibrate at a characteristic frequency. They then emit photons with the same wavelength that can be measured to an accuracy of one part in 10^{14}, which corresponds to a time accuracy of one second in every million years. This definition was carefully chosen to closely match the older definition of one second, which was calculated from the Earth's daily rotation. Even more recently, this definition had to be re-standardised because of the consequences of Einstein's theory of general relativity, which implies that the rate at which time passes depends on the gravitational field as well as the relative speed at which you are travelling. Our definition of time refers to a caesium atom at rest, at sea level and at a temperature of zero Kelvin (-273 degrees centigrade).

Age of the Earth

So how are things actually dated that existed before the written record? For physical items on Earth, we can analyse their chemical composition and use "radiometric dating" to determine their age. In a nutshell, the relative numbers of certain atoms change in a well-defined way over time and by counting the number of mutated atoms we can date the object. This is such an important tool of measurement that we should look at how it works in practice.

Nothing lasts forever. This will become a familiar phrase in this book, and one that applies to individual particles and atoms too. Atoms are complicated things, clouds of electrons surrounding a nucleus of protons and neutrons that themselves have unique properties and an internal structure consisting of yet smaller particles called quarks. Electrons are thought to be elementary particles: they do not have any internal structure. At school it is taught (for simplicity) that electrons orbit the nuclei of protons and neutrons, rather like a planetary system orbiting the Sun. At university one learns that this is actually not true and that the particles act like tiny packets of energy that occupy some fuzzy allowed region. One can never know precisely where they are.

You might think that any given atom could, well, just be an atom forever. But many atoms are characterised by highly unstable configurations. Moreover, their internal structure can change quite suddenly, albeit following very well defined physical rules derived from our theories of particle physics. An unstable atom can move down the periodic table into a new form by radioactively decaying and emitting a high-energy charged particle. This is known as ionizing radiation and it can be deadly, causing genetic mutation and cancer. Unstable atoms will usually try to evolve into something that is more stable, such as iron or lead. Not all atoms are unstable. In fact, from our knowledge of particle physics, the first 80 elements of the periodic table should have isotopes that are stable.

It is not known what triggers the process of radioactive decay. Therefore, we cannot predict when a given atom will change its internal structure. It seems to be completely random, just as demanded by our standard interpretation of quantum mechanics. Yet it may only appear random because we do not know the physical cause and effect—the hidden variables at work. So how can we use a random process as a measure of time? A collection of atoms of the same type will decay on the same average timescale, even though some individual atoms will last longer than others. Consequently, by looking at lumps of material containing trillions of atoms, we can accurately calculate this average time, known as its half-life, the time interval over which half of the atoms have decayed and mutated.

The most unstable atoms decay in a fraction of a second; others are stable for a timescale longer than the age of the universe. Atoms that have been used for radiometric dating decay on timescales from 10 years (e.g. tritium) to over 100 billion years (e.g. samarium). Atoms differ in their exact composition of protons, neutrons and electrons. Consider uranium-238, so called because it has 146 neutrons and 92 protons. It has a half life of 1.41×10^{17} seconds (about 4.5 billion years), meaning that there is a 50:50 chance of a single atom changing to lead within this timescale. If a rock containing this form of uranium formed 4.5 billion years ago, today half of the atoms will have turned to lead and half will remain unchanged. Four-and-a-half billion years into the future, one-half of these remaining uranium atoms will have also mutated. In other words, three quarters of the original uranium atoms will have become lead. By counting the ratio of lead atoms to uranium atoms, we can determine the age of the rock. The mineral zircon is typically used to date the oldest rocks on Earth since as it forms, it incorporates uranium into its crystalline structure but not lead. Any lead atoms found in zircon today must have originated from the decay of the uranium atoms that were trapped inside its structure when it crystallised and formed.

The British scientist Ernest Rutherford discovered these processes and in the early 1900s used them to measure the age of rocks. His work led to the remarkable finding that the oldest rocks on the Earth's surface formed 4.4 billion years ago, so we know that the planet is at least as old as that. Employing the same techniques, we can date fossils, skeletons, stone-age tools and even meteorites, and we can reconstruct the history of our species and solar system. A century ago it became clear that our short lives are just an infinitesimal moment in the passage of history and time in the eyes of the Earth.

Hundreds of tons of matter from meteoroids land on Earth every day. Most of it falls to the ground as dust, but about a tonne of this material lands as pebble-sized objects or larger. These space rocks are leftover debris from the formation of the planets. Radiometric dating shows that they formed around 4.54 billion years ago. This is slightly older than the rocks on Earth, which is as we expect since it took the Earth about a hundred million years to cool down sufficiently so that its rocky crust could form.

What about dating the age of the other planets? Of the Sun? Of the universe itself?

Moon rocks brought back in the 1960s by the Apollo astronauts have been analysed and show a very similar composition and age to the oldest rocks on Earth. We have not yet brought back anything from other planets, but we are fortunate to already have some pieces of Mars to analyse right here on Earth. At various times across the history of our solar system, giant asteroids collided with Mars with such violence that they scattered debris from its surface into orbit around the Sun. Some of those pieces have since fallen to Earth as meteorites, and we find that many are just as old as the rocks on Earth. We believe that they originated on Mars since they have exactly the same isotopic abundances of certain elements that we observe in its atmosphere, which is very different from our own atmosphere.

Scientists do not like to rely on one single method or experiment—they like to cross-check their results. So we look to see whether there are methods for calculating ages other than radiometric dating. For example, the growth rings of trees follow a yearly cycle. By matching living trees to fossilised trees, we can trace both the climate conditions and the age of the trees back over at least 10,000 years. Sedimentary or algae deposits in lakes also follow yearly cycles and can be used to crudely measure the age of rock deposits over the past million years. These techniques agree perfectly with radiometric dating techniques. However, the most impressive test of radiometric age dating is a comparison of the age of the Earth with the age of the solar system that we can calculate from our understanding of how the Sun has evolved over time.

Ages of the Stars

We do not have the technology to extract any material from the Sun and bring it home to analyse in the laboratory. Even if we could, it is not obvious how we could use it to accurately measure its age. So how can we determine the age of our star the Sun, something we cannot touch? Stars shine and evolve according to physical principles that we know and understand very well. In fact, we understand them to such an extent that measuring a star's surface temperature, its luminosity and its mass is enough to determine how long it has existed as a star. We can also predict how it will evolve in the future and when it will eventually end its life.

Even stars do not last forever. Eventually, they burn all their fuel and die, some exploding as spectacular supernovae and others simply fading to darkness. The lifetime of a star depends on how massive it is, which determines the rate at which it generates energy from nuclear fusion at its core. Stars of the same initial mass and chemical composition will evolve in exactly the same way, but more massive stars burn their fuel faster, leading to higher temperatures and shorter lives. By measuring the present temperature and brightness of a star, we can pinpoint its particular evolutionary stage and hence its age by comparing it to our theoretical calculations of stellar evolution.

We can therefore measure age by modelling the change of something over time and then matching the model to fit current observations. This only works if we know all the important physical processes that may be at work, and history shows that mistakes can be made. For example, the first (incorrect) estimates of the age of the Earth were made by assuming that it was very hot when it formed and that it slowly cooled down by radiating heat until it reached the temperature it has today. The inside of the Earth is obviously very hot, as anyone witnessing the eruption of a volcano can attest. So, if the planet started out as a molten rock, it would have slowly radiated infrared photons (heat energy) and cooled down to its current mean surface temperature. All we need to know is how quickly a lump of hot rock radiates away its energy, which can be measured in a laboratory or calculated theoretically.

The nineteenth century gave rise to intense debates on the age of the Earth. William Thomson (Lord Kelvin), originator of the fundamental temperature scale, used this technique to estimate its age and came up with a timescale that was just tens of millions of years. When the same ideas and principles were used to calculate the age of the Sun, a similar timescale was found and everything seemed to be consistent. However, the debate continued between physicists and geologists and biologists who were studying processes on our planet which seemed to take place over hundreds of millions of years. Although the physicists made a correct calculation based on the known physics of the time, the estimates were wrong since there were additional unknown heat sources at work from physical processes that had yet to be discovered. In the case of the Earth, scientists were not aware of the heat that radioactivity could supply until the work of Rutherford. The Earth contains a small fraction of those unstable elements, like uranium-238, which decay on a several-billion-year timescale, releasing energetic particles that are absorbed by the surrounding rocks and thus increasing their temperature. Volcanoes and lava flows, plate tectonics and thermal hot springs are all powered mainly by radioactivity within the Earth's interior. By not including this heat source within their models of a cooling Earth, scientists had obtained the incorrect answer for its age.

Rutherford recalls presenting his results in a 1904 speech to the Royal Institution and observing Lord Kelvin in the audience: "I came into the room, which was half-dark, and presently spotted Lord Kelvin in the audience and realised that I was in for trouble at the last part of my speech dealing with the age of the earth, where my views conflicted with his. To my relief, Kelvin fell fast asleep, but as I came to the important point, I saw the old bird sit up, open an eye and cock a baleful glance at me. Then a sudden inspiration came, and I said Lord Kelvin had limited the age of the earth, provided no new source [of heat] was discovered. That prophetic utterance refers to what we are now considering tonight, radium! Behold! the old boy beamed upon me."

Likewise, determining the age of the Sun correctly could not be achieved until the early twentieth century following the incredible discovery of Einstein that you could take a little piece of matter and turn it into a vast amount of energy. Soon after this, the British astrophysicist Arthur Eddington realised that the temperature and pressure at the centre of the Sun may be so high that hydrogen atoms could literally fuse together, forming helium and fuelling the Sun via the conversion of a tiny amount of their mass into energy.

This is the principle behind the hydrogen fusion bomb, such as the Russian Tsar Bomba. This was the most destructive weapon ever made by our species and was exploded in 1961 on a remote island off the northern coast of Siberia. It was equivalent to detonating 50 million tonnes of the explosive TNT (which would fill a cube 300 metres on each side) and was over a thousand times the combined power of the two nuclear fission bombs used to destroy Hiroshima and Nagasaki in the Second World War. The energy yield of nuclear fusion is immense; the Tsar Bomba released the same amount of energy as a Richter magnitude nine earthquake. If you could contain the energy from this single bomb and convert it into electricity, it would have been enough to power Switzerland for a year. Nuclear fusion is a very clean energy source since it does not use or release radioactive elements. But so far scientists have been unable to maintain a controlled and sustained fusion reaction.

At the centre of the Sun, the equivalent of a trillion hydrogen bombs are continuously detonating each second, providing the energy that makes the Sun shine. Just the radiation from the Sun that lands on the surface of the Earth each second has the energy of the Tsar Bomba! If we could cover an area in the Sahara desert that is about the size of England with modern solar panels, we could provide the entire planet with all of its energy needs. Once we take this new energy source into consideration in our calculations, we realise that the Sun has an enormous reservoir of fuel. Moreover, we have calculated that it has been shining for 4.56 billion years. That is very close to the age of the oldest rocks on Earth, which is not a coincidence since we believe that the Sun and the planets must have formed at the same time.

So now let us get back to the age of something even grander—our universe. We can apply our knowledge of how stars evolve to determine the age of our galaxy. And by looking for the oldest stars, we can estimate a lower limit of the age of the universe in which it lives. When the Sun eventually burns all of its fuel, it will shed its outer layers in a majestic final "planetary nebula" phase, and the inner regions will collapse into a dense core known as a white dwarf. But there is no immediate concern. The Sun has enough fuel to continue shining for another seven billion years.

I will explain more about how stars work and how they evolve and eventually die in a later chapter. All we need to know for now is that the lifetime of a star depends on its mass. A star that is twice as massive as the Sun will live for only 800 million years. Conversely, one that is half the mass will last for 20 billion years. How do we use this knowledge to estimate the age of our galaxy? Well, it turns out that stars often form together in large groups or clusters, such as the beautiful Pleiades star cluster often called the Seven Sisters. Most people can see six or seven stars with the naked eye, although the cluster actually contains hundreds of fainter stars. Take a look one clear night—it is a good test of your eyesight. Pleiades is the brightest visible star cluster. You can see it in the constellation of Taurus, close to Orion (the giant hunter holding a club). Within these clusters the stars all form at the same time but with a range of different masses. Depending on the age at which the stellar cluster forms, all the stars above a certain mass will have died and faded out of sight. The most massive stars found in the oldest star clusters of our galaxy are measured to have about 70 percent of the Sun's mass, which means that these stars are over ten billion years old! The Milky Way galaxy is therefore over twice the age of the Sun and the planets.

The universe must therefore be at least this old. But, we are still left with the question, Is the universe infinitely old or did it appear a finite time ago, a time that can be measured? The answer was made possible because of several amazing discoveries in the 1920s that revealed the nature and true extent of the universe. The story begins with the measurement of the distances to the stars and then the discovery of distant galaxies beyond our own, culminating in the realisation that the universe was expanding, which in turn implied that it had a beginning.

Distance to the Stars

Distances on Earth are straightforward to measure—we can use a standard ruler to define length. But what about measuring the distances to the Sun or to the faraway stars that we have not visited. . .yet? We can achieve this using simple geometry invented by the ancient Greeks, who defined space mathematically so that they could quantify and measure it. The unit of length at that time was a "stadion", the length of a Greek stadium, which was about 200 metres. In 1668 the English philosopher and one of the founding members of the Royal Society, John Wilkins, proposed a decimal based standard unit of length. His inspiration was the architect and astronomer Christopher Wren's proposal of defining a standard distance measure as the length of a simple pendulum that swings once from left to right in one second. The alternate suggestion at the time was to define the metre as one ten-millionth of the distance from the equator to the North Pole. I know which method I would have chosen! However, the French academy of sciences selected the latter method in 1791, arguing that the force of gravity varies slightly depending on where you stand on Earth's surface. This is because the Earth is not an exact sphere. It is slightly wider at the equator due to its rotation—the centrifugal force causes the distortion because the Earth is not a solid object but mainly molten "liquid" rock. Therefore, the pendulum would take slightly longer to swing if you stood on the equator where the Earth's gravitational force is weaker. Indeed, on the equator your weight would be half a percent less than it would be at the poles.

This distance to the North Pole was carefully determined. A standard prototype metre length was established, although it was short by a fifth of a millimetre due to the miscalculation of the shape of the Earth. Thankfully, we now have much more accurate techniques by which we can measure length. Since 1983 one metre has been defined as "the distance that a photon of light travels in a vacuum in 1/299,792,458 of a second". As for the modern standard for determining time, this measure of length is intended to be as close to the older definitions as possible. The difference is that time and length are now uniquely specified in terms of a quantity that never varies.

Aristarchus (300 BC), of the Greek island Samos, was the first to measure the distance from the Earth to the Sun and from the Earth to the Moon as well as their sizes. Isn't that remarkable? Over two thousand years ago, long before telescopes and calculators, Aristarchus determined our place in the solar system. He made the correct premise that the Moon reflected the light from the Sun and that when the Moon is exactly half illuminated, it must form a right-angled triangle with the Earth and the Sun. By measuring the angle between the Sun and the Moon, he could use trigonometry to calculate the relative distances from the Sun to the Moon in terms of their relative sizes. Measuring this angle was difficult since the Sun is so bright and the angle is about 89.8 degrees—Aristarchus estimated it to be 87 degrees. Then he found the relative distance from the Earth to the Moon in terms of the ratio of their sizes using another geometrical configuration, an eclipse of the Moon by the Earth. To calculate the distances in terms of an absolute length, stadions, he "simply" had to insert the known size of the Earth. This was an incredible achievement and was the foundation of the "distance ladder" that enables us to bridge the gaps in measurement techniques to ever more distant objects in the universe.

Aristarchus also realised that the stars in the night sky must be much more distant than our Sun since they showed no detectable parallax. Parallax is a technique which uses the change in apparent position of an object viewed from two different places to measure its distance. This is the reason why most animals evolved to have two eyes that point in similar directions; it is a good advantage to have over your enemies to be able to judge their distance away from you. Pigeons do not have this immediate ability. Each eye sees a completely different field of view since their eyes are on opposite sides of their heads. One proposed explanation of why pigeons bob their heads up and down as they are moving is so they can measure distance by parallax. To see how it works, hold up your finger and look at how the position of your finger changes relative to distant objects as you look with each eye. Now move your finger twice as close and the distant object appears to move twice as far from side to side. This change in apparent position can be measured as an angle. Then all you need to know is the separation of your eyes, and you can measure the distance to your finger using geometry. The further apart your eyes are, the more accurately you can measure the angle.

In the nineteenth century, the quality of telescopes had improved to such an extent that the German mathematician and astronomer Friedrich Bessel was able to measure the parallax of stars beyond the Sun and thus their distances from Earth. He observed one particular star called 61 Cygni and measured the change in its apparent position between January and June as the Earth moved halfway around the Sun. This is the furthest apart that Bessel could put his telescope "eyes". He found that the distance from Earth to 61 Cygni was 658,000 times the distance from the Earth to the Sun (10.4 light years away!). The apparent universe was already becoming very large. But because of the limitations of the telescopes of the time, parallax could only be used to measure the distances to the nearest stars. The true extent of our galaxy was not known until 1908 when the American astronomer Henrietta Leavitt discovered a way of measuring the distances to much more remote stars, revealing that our galaxy was at least 50,000 light years across. This technique (which I shall describe shortly) also enabled the discovery that our galaxy is not alone in the universe.

A Vast Universe

How can we measure the size and age of the universe more precisely? The idea is quite simple. But the concepts are profound, and its measurement represents the pinnacle of cosmological theory and observation. The story begins in 1919, shortly after Einstein published his research on general relativity, when the American astronomer Edwin Hubble made the discovery that there was a lot more to the universe than just the stars we can see with our naked eyes.

Prior to Hubble's discovery, the known universe consisted only of the stars within our Milky Way galaxy. Nothing else was known to lie beyond it. Hubble changed this view by measuring the distances to some of the fuzzy nebulae that dotted the night sky. Nebulae were noted to be extended blurred patches of light, neither starlike nor cometary in origin. The first mention of these strange objects was made in 964 by the Persian astronomer Abd al-Rahman al-Sufi, who noticed a faint extended patch of light where we now know the Andromeda galaxy to be located. For a thousand years their nature was unknown until Hubble used the technique pioneered by Leavitt to measure the distances to stars within the nebulae.

In the early 1900s women were not allowed to operate telescopes, but Leavitt and several other women were hired to perform the rather dull task of measuring and cataloguing the brightness of thousands of stars on photographic plates. They were known as the "human computers". When Leavitt compared the brightness of the same stars on images taken on different nights, she noticed that they varied in a well-defined and periodic way—the luminosity of the brighter stars were changing over a longer period of time than the fainter stars. The stars she was studying are called Cepheid Variables; they pulsate on a timescale from one day to a couple of months, leading to a regular change in their brightness. Their regular pulsation is a consequence of the escaping radiation that causes the outer regions of the star to expand. As it expands, it cools down and contracts, and the process begins again. Leavitt published her results in 1908, and Hubble himself wrote that she deserved the Nobel Prize for her work. Unfortunately, she died before the importance of her discoveries was recognised.

Imagine a light source that is always the same strength. Astronomers call these standard candles and use them to measure distances to faraway objects. For example, if all stars had the same brightness, more distant stars would always appear fainter. This is because the number of photons we receive diminishes as one over the square of its distance, just like the force of gravity. A star that was four times fainter than another would imply that it was twice as far away.

In 1698, the Dutch astronomer and physicist Christiaan Huygens estimated the distance to the star Sirius by comparing its brightness with that of the Sun. It was hard to measure the equivalent brightness of the Sun. But by using a pinhole to reduce the amount of sunlight until it appeared as bright as Sirius, Huygens estimated that Sirius must be 27,000 times as distant as the Earth–Sun distance. Since Sirius is the brightest star, it was thought it must be the closest. However, the assumption that all stars have the same brightness (that they are all standard candles) was incorrect. In fact, Sirius is much brighter than our Sun—about 25 times brighter—so Huygens's distance estimate was short by about a factor of five.

Any Cepheid Variable star whose brightness changes with the same period will have the same total luminosity. Therefore, just by measuring the pulsation period and apparent brightness of a Cepheid Variable star, we can determine its distance. This technique relies on knowing the absolute distance to at least one nearby Cepheid Variable, which can be calculated, for example, by using parallax. We will see later how supernovae can also be used as standard candles to measure distance across most of the visible universe, which will take us to the recent discovery of dark energy.

Cepheid stars are bright enough to be observed in some of the fuzzy nebula. Hubble measured their pulsation period and apparent brightness using the new one-hundred-inch telescope at Mt Palomar in California. Using this next step in the distance ladder, in 1925 Hubble found that the Andromeda nebula was at least ten times further away than the most distant stars in our galaxy. It was another profound realisation that the universe was a truly vast place containing other "island universes"—galaxies like our own each containing billions of stars. But there was more to come.

While astronomers were still developing the tools and techniques needed to measure the universe, the theoretical cosmologists were busy studying Einstein's equations and attempting to apply them to the universe. They started with the assumption that space is isotropic (similar in all directions) and that the universe was homogeneous (similar in all places). This is the Copernican principle that I mentioned earlier as being central to modern cosmology. Several brilliant theorists then independently came up with solutions to Einstein's equations of general relativity in an expanding or contracting universe. The first to do this (in 1922) was the Russian physicist Alexander Friedmann, who derived equations that related the size of the universe as a function of time to its (then unknown) matter and energy content. It was an amazing feat of human intellect to be able to speculate about such radical ideas. The results of the work of Russian scientists took time to percolate and be translated into English. In the meantime, several Western physicists came up with the same conclusions, including Georges Lemaître, about whose work Einstein commented: "Your math is correct, but your physics is abominable". Einstein disliked the concept of an expanding or contracting universe.

The Universe Is Expanding

How do we measure the speed at which a star or a distant galaxy moves? And with respect to what do we measure its movement? Usually we measure speed relative to the Earth's surface, which is at rest for us. But our real motion through space is quite complicated. Standing on the equator, you would have an effective speed of 500 metres per second due to the Earth's rotation. The Earth itself is moving around the Sun at 30 kilometres per second; the Sun and the entire solar system are moving around the centre of the galaxy at 220 kilometres per second. It is a good thing that all that spinning around does not make us dizzy!

The motions of distant objects can be measured using the "Doppler effect". This is similar to the way the noise of a passing motorcycle changes from a higher to a lower frequency sound as the sound waves are first compressed due to their motion towards us and then stretched as they move away. The change in frequency (or wavelength) is proportional to the speed at which the motorcycle is travelling. Light behaves in the same way: By reflecting microwaves from oncoming motor cars, traffic police can measure the frequency change of the reflected photons and thus measure their speed. If you were to run towards me with a red flashlight, the photons would appear to me as being slightly shifted towards the blue part of the colour spectrum as the wavelengths are compressed. To the person running with the flashlight, the photons would appear unchanged as red, just as the sound of the motorcycle does not change for the rider. To change red light to blue light you would need to run quite fast. . .at around one thousand kilometres per second.

The conceptual difficulty that often arises is that the speed at which the photons reach me is unchanged. They always travel at the same speed. That is one of Einstein's fundamental postulates. It does not matter how fast you are moving; those photons will always travel away from you at the speed of light. You do not get to add the velocity at which you are moving. It seems paradoxical, but it isn't. Relativity is full of apparent paradoxes that are always resolved once the question is stated carefully and the appropriate calculation is made.

The speed at which another star or galaxy is moving towards or away from us can therefore be measured by observing the spectrum of its light. If it is moving away from us, then the wavelengths of all the photons that it emits will be stretched and lengthened a little. The spectrum of light from a galaxy is measured in the same way that a prism splits sunlight into the colour spectrum. Certain elements on the surfaces of stars emit strong radiation at specific frequencies (or colours), called emission lines. The idea is to measure the increase (red shift) or decrease (blue shift) of the wavelengths of the emission lines compared with the spectral lines from the same element at rest in a laboratory on Earth. The change in wavelength is proportional to the speed of the galaxy. In 1912 the astronomer Vesto Slipher was the first to use this technique to show that the nearby galaxies were indeed moving very fast and that most were moving away from our own galaxy at thousands of kilometres per second.

In 1927 Lemaître made the first connection between theoretical models and observations to argue that the universe was expanding. Using the distances and velocities of the nearby galaxies measured by Hubble and Slipher, he showed that they were moving away from our own Milky Way and that the most distant galaxies were moving away faster. This is exactly what you would expect if, for some reason, the space between all the galaxies were expanding: Lemaître, a Belgian astronomer and priest, had discovered the big bang! He published his work in French in a relatively obscure journal, which is perhaps why Hubble is generally credited for this discovery which he made independently just two years later. In 1929 Hubble published his famous paper, "A Relation between Distance and Radial Velocity among Extra-Galactic Nebulae".

Prior to these discoveries by Lemaître and Hubble, it seemed natural to conclude that the universe did not change on very long timescales. The incessant motion of the Earth around the Sun, which appeared to be the same each day, and the stars that appeared in the same place each night, all gave the impression that the universe was static and unchanging. These results completely changed our perception of the universe. It was not a static timeless entity. Rather, it was rapidly expanding, which implied that it was smaller in the past and ultimately originated from a small and very dense region—that it had a beginning.

The rate at which space is expanding today is called Hubble's constant. Its value is about 25 kilometres per second per million light years of space with an accuracy that is better than ±one kilometres per second. This means that a galaxy which is 100 million light years away should be moving away from us at 2,500 kilometres per second, and one that is twice as distant would be moving twice as fast. The reason that more distant objects appear to be moving away from us faster is because there is more space between us, and it is the space that is expanding uniformly. It is rather like stretching an elastic band. Take one and stretch it, and notice how fast your fingers are moving apart compared to a point at the middle and one of your fingers; it is twice as fast since there is twice as much space between your fingers. That is why Hubble's constant is quoted in terms of speed per unit of distance. Since speed itself is distance per unit time, the distance units cancel out, leaving the overall unit of Hubble's constant as one over time. If we calculate one divided by Hubble's constant, we find a value of 13 billion years.

This measurement of time should be close to the age of the universe. It is an estimate of the time it has taken for the galaxies to have expanded into their current positions starting from a single point in space. We can now make an estimate of the size of our visible universe by calculating the furthest distance that light could have travelled in this time—that is simply its age multiplied by the speed of light. We cannot see further than that since there has not been enough time for light to reach us from more distant regions. And photons, well, they travel at the speed of light. That gives us a rough estimate of the size of our visible universe that is tens of billions of light years across. Our universe is indeed old and very large. It could be much larger, even infinitely large. But since we cannot observe those more distant reaches of the universe from which light has not had time to travel to us, we believe that we can never know its true size.

Early estimates of the Hubble constant were difficult to make, which motivated the construction of the Hubble Space Telescope. Its main purpose was to accurately measure the Cepheid Variable luminosity relation in distant galaxies. By the time the space telescope completed this task, various other observational data and techniques had already achieved a similar accuracy, confirming its most recent estimate. The list of scientific achievements that have come from the use of the space telescope is long, culminating most recently with the Hubble Ultra Deep Field,[1] the longest exposure image of the universe ever taken. It verifies the Copernican principle and reveals galaxies which had already formed over ten billion years ago.

We now have the tools and data necessary to determine a very precise age of the universe. The basic idea is to calculate how long it has taken since the Big Bang for the galaxies to expand into their current configuration. In detail this depends on the composition of the universe. Once we know what the universe is made of and how much mass and energy there is in all the different forms that they can take, we can calculate its past history and also its future using the mathematical equations

---

[1] http://www.nasa.gov/vision/universe/starsgalaxies/hubble_UDF.html

derived by Friedmann and Lemaître. It is only in the last decade that our knowledge of the cosmic inventory has allowed us to accurately determine the age of the universe. Although we still struggle to understand its components in detail, the nature of the mass and energy does not affect the conclusions. It is just the total amount that is important and not the details of the forms. If we put in everything we know and are honest about all the uncertainties, we can calculate the age of the universe to be 13.8 billion years plus or minus 50 million years.

The Cosmic Inventory of Matter

This leads us to the final topic of this chapter, how do we measure the mass of the universe and what forms does it take? We would like to determine whether or not there is enough matter in the universe to reverse its expansion and to find out whether it is made of the same basic stuff as us. . .atoms.

Our standard unit of mass is the kilogram—the only international scientific unit that is still defined by an actual physical object. For most practical purposes one kilogram is actually very close to the weight of one litre of water (that is correct to one part in one hundred thousand). The "official kilogram" is a small cylinder (to minimize its surface area against corrosion) of height and diameter 39.17 millimetres, kept in a vault in Paris and requiring three independent keys to open. It is made of 90 percent platinum (which is dense and quite stable) and 10 percent iridium (which increases its hardness). Official copies are given to other countries to serve as their national standards. However, it was recently found that the standard kilogram is slowly getting more massive as it adsorbs molecules from the air. Consequently, scientists are working on a new standard of mass that is based on an invariant measure, such as a fixed number of atoms of carbon.

In order to estimate the masses of galaxies and the number of atoms in the universe, it was first necessary to calculate accurate masses of the stars, since stars seem to be the most important component of galaxies. Whilst Aristarchus had estimated the size and distance to the Sun using geometry, geometry alone does not allow the determination of the mass of something. Accordingly, his methods were not sufficiently accurate for our needs today. The mass of our Sun was first measured in the nineteenth century, two hundred years after Isaac Newton derived the equations of planetary motion which link together the distance from the Earth to the Sun, the orbital period of the Earth (one year) and the unknown mass of the Sun.

The first step towards determining a precise mass of the Sun was to make an accurate measurement of its distance from Earth. This was achieved using parallax during the transit of Venus across the Sun's surface in 1761. The transits of Venus occur in pairs separated by about 8 years every 110 years. In 1716 the English astronomer Edmond Halley proposed that the next events should be observed at different places across the planet and the time of the transit accurately recorded. This way, the change in the apparent position of Venus against the Sun could be determined from different viewpoints on Earth all at the same time. He died

20 years before it took place. But in one of the first international collaborative experiments, astronomers from across the world set off on long journeys to observe the transit event from widely separated places across the planet.

The tale of French astronomer Guillaume Le Gentil is particularly sad. In 1760 he set sail to India to observe the transit. On the way, however, he learned that a war had begun between France and Britain and thus it was too dangerous to try to reach his destination. He was a determined fellow, so he then headed to an alternative part of India only to find that it was also occupied by the British. His ship was ordered to turn back. Although the sky was clear, when at sea it was impossible to observe the transit because of the motions of the boat. Disappointed, he decided to await the next transit event eight years later since he had already travelled so far. He sailed to Manilla in the Philippines to observe the 1769 transit but was turned back by the Spanish authorities there. Unperturbed, he headed back to his original destination in India which was back under French control. He constructed his observatory and waited patiently. Even though every day of the previous weeks had been clear, on the day of the event the skies clouded over and he was unable to see anything. He only just managed to avoid sinking into insanity and started the long journey home to France, with many adventures along the way. He eventually arrived back in Paris 11 years after he had set off to find that he had been declared legally dead, his wife had remarried and his relatives had shared his estate amongst themselves. . .What some people will do in the pursuit of scientific discovery!

The other expeditions were a lot more successful, and the distance to the Sun was finally measured using parallax and trigonometry. But there was still a missing piece of information: the constant of proportionality in Newton's law of gravity, which tells us exactly how much force is exerted by a given mass of material. Newton called it G, the gravitational force constant. The final step in measuring the mass of the Sun had to wait until the British scientist Henry Cavendish calculated G in 1798 by measuring the attractive force between two 12 inch lead spheres each weighing 158 kilograms. The force between them was tiny, equivalent to measuring the weight of a single grain of sand. The spheres attracted each other and moved a torsion balance just one tenth of an inch, from which Cavendish could calculate G—a remarkable experiment to accomplish at the time. Putting all the numbers into Newton's equations, we find that our yearly motion around the Sun at a distance of 150 million kilometres requires the Sun to have a mass of 2.0×10^{30} kilograms. That is a little more than 1,000 times the mass of Jupiter and 330,000 times the mass of the Earth.

Now that we know the mass of the Sun, and since most of the stars in our galaxy are similar to our Sun, we can simply add up the number of stars to make a simple estimate for its total mass. Careful counting by computers using deep imaging data from our telescopes reveals that there are at least 100 billion stars in our galaxy! We can then obtain a rough lower limit to the mass of the universe by adding up all the stars within all the galaxies that we can see. That is about 100 billion galaxies each containing an average of a hundred billion stars! Whilst this is an enormous mass, the volume of the universe is large. If we could spread all the observed stars and gas out uniformly over space, any given cubic metre would contain less than one hydrogen atom. We can calculate just how much mass is needed for gravity to halt and perhaps reverse the cosmic expansion. We find that we would need the equivalent of almost ten hydrogen atoms per cubic metre of universe.

When I started doing research in astrophysics in the 1990s, a strong theoretical bias existed among many cosmologists who thought that there must be enough mass in the universe to exactly stop its expansion. From the equations of Einstein and Friedmann this would imply that the total energy of the universe was exactly zero, such that the positive energy in the expansion was exactly balanced by the negative attractive force of all the matter. It was interesting to watch the subject develop—as hard as my colleagues searched, they could not come up with enough mass, implying that the universe had a net positive energy and would expand forever. Could there be another component of matter unlike normal atomic matter which could contribute to the mass of the universe and that, for some reason, we cannot directly see? The bizarre answer was yes. It was a discovery made over 80 years ago by the Swiss astronomer Fritz Zwicky.

Dark Matter

Gravity relentlessly strives to pull matter together and is the force responsible for creating an entire hierarchy of structure in the universe, from individual stars to the giant clusters of hundreds of galaxies. Galaxy clusters are the largest systems to have emerged from the gravitationally driven structure-formation process. They are regions of the universe that contained just enough matter to reverse the cosmic expansion, resulting in a gravitational collapse and the formation of an equilibrium configuration of galaxies. In 1933 Zwicky made the first measurement of the mass of one of the largest known cosmic structures, the Coma galaxy cluster, and found a rather unexpected result—one that is still not explained to this day.

To measure its mass, he used a very neat little formula that was first derived to study the properties of gases. As an analogy, consider that the size of a balloon depends on the temperature and the number of molecules that you blow into it. Temperature is just a measure of how fast those molecules are moving. The movement of the molecules gives rise to a pressure force as they collide with the inside surface of the balloon, thus preventing it from collapsing. In a galaxy cluster it is the random motions of the individual galaxies that prevent them from collapsing together under their mutually attracting gravitational forces.

Using the Doppler technique to measure their velocities from their light spectra, Zwicky found that the galaxies are whizzing about in the Coma cluster at thousands of kilometres per second, from which he derived its mass to be equivalent to 10^{15} Suns. When he actually counted the galaxies and estimated the total number of stars, he found a number that was one hundred times smaller. There simply was not enough mass in the observed stars and galaxies to hold the cluster together. To prevent the galaxies in the cluster from dispersing and simply moving away from each other, he speculated that there must be an enormous amount of mass in the cluster that he could not see. He called it *dunkle Materie*, or dark matter.

Three more decades passed before further evidence for this "missing matter" was found, this time within individual galaxies. The stars in spiral galaxies, like our Milky Way, live in a flattened disc-like structure. They move on nearly circular orbits about the galactic centre. In the same way as the mass of our Sun can be measured using the Earth's orbital period and its distance from the Sun, we can calculate the mass of an entire galaxy by measuring the orbital speed of the stars as far away from the centre as possible. We can even reconstruct how the mass is distributed by measuring the rotational speeds of stars at different radii. Since most of the stars in a galaxy are concentrated within their inner regions, it was thought that the more distant stars should be moving more slowly than the inner stars, just like the planets orbiting the Sun; Mercury completes its orbit in just 88 days, whereas Neptune takes over 60,000 days. The reason for this is that the gravitational force from the Sun is much weaker at the distance of Neptune. However, in 1975 the American astronomer Vera Rubin presented the surprising result that the speed at which the stars rotate about the centres of their galaxies is almost constant with distance, nothing like the expected decline with distance. This result implied that galaxies also contain a large component of matter that was not readily visible—dark matter that stretched at least as far as the outermost stars to enable their high orbital speeds.

Today we know a lot more about the quantity of dark matter and how it is distributed within and around galaxies and galaxy clusters. One of the most powerful and convincing measurements can be made using gravitational lensing, which relies on the link between gravity, space and time that was made by Einstein. Gravity literally curves the fabric of space, implying that photons of light do not follow simple straight lines through the universe. Every photon that we see from a remote star in a distant galaxy has travelled across the universe on its own roller-coaster journey to get here, curving around every massive object in its path. The deflection of light by massive objects is very similar to the bending of light by a magnifying glass; more highly curved lenses distort the images more strongly. In an analogous way, the amount of distortion in the images of distant galaxies that happen to lie behind a galaxy cluster can be used to measure the curvature of space due to gravity and hence the cluster's mass. The beauty of gravitational lensing is that it can be used to measure the mass distribution beyond the edge of the visible sizes of galaxies where their stars live. In fact, we now know that the sizes of the dark matter haloes surrounding galaxies are at least ten times as large as the region within which the stars orbit.

Literally thousands of research papers have been published on dark matter, and its existence, quantity and distribution in space are well established. I have written research papers on dark matter, many of my students have written papers on it and my student's students have written papers on it. However, despite the efforts of numerous scientists around the world, its precise nature still remains a mystery. Over 80 years have passed since Zwicky identified a component of matter that is unlike anything with which we are familiar. It is certainly not in the form of atomic material, since those particles would easily be detectable: they would radiate photons that could be observed. We have found that the atomic material that constitutes us, as well as planets and stars, is just a minor part of the universe; there is about five times as much dark matter as atomic matter.

That is another profound thought: Even the atoms of which we are made are only a small fraction of all the matter that exists in the universe! But the conclusion that follows from our cosmic inventory is perhaps the most thought-provoking. If you add up all the matter in the universe, in all the dark matter, atomic matter, neutrinos and everything else that attracts by the force of gravity, you do not find enough mass in the universe to prevent its expansion. Gravity will fail in its task to assemble ever larger and more complex structures, and the universe will expand and grow in size for eternity.

We know what dark matter is not and we have some ideas as to what it could be. It is thought to be composed of vast numbers of a new fundamental particle that has yet to be detected, a particle that does not radiate light, which is why we cannot simply see it. All the data we have suggests that all galaxies live at the centre of a huge extended distribution of these particles that is ten times as massive as all the visible stars. One candidate is a "neutralino", a hypothetical particle that theoretical particle physicists have predicted to exist to explain certain features of the fundamental forces of nature. There is a fairly long list of other possible particles that have been proposed as possible dark matter candidates, some of which go by the names of axions, neutrinos or gravitinos.

As our solar system orbits around the galaxy, we think that there are literally billions of dark matter particles passing through our bodies every second of every day. We cannot feel anything since these particles must interact very weakly with normal atomic matter, which is why we sometimes call them WIMPS—weakly interacting massive particles. Such a dark matter candidate would rarely collide with an atom because it would need to pass extremely close to the nucleus. But now and again this will happen, leaving open the possibility that it can be detected with very sensitive laboratory experiments here on Earth. Several experiments around the world are attempting to detect the dark matter particle by looking for these collisions, but the technological challenge is immense and much harder than trying to find a needle in the proverbial haystack. The experiments have to detect the resulting motion of a single atom within a lump of material made up of about 10^{30} atoms! If that does not sound difficult enough, other particles can also mimic the same signals as dark matter particles, for example the ionising radiation from radioactive decays of unstable elements in the walls and containers of the laboratory.

Although a signal has yet to be found, the cryogenic dark matter search is one of the most sensitive experiments in operation. It is designed to measure the sound waves that occur when a collision takes place between a WIMP and a single atom within a germanium crystal. The experiment is housed deep underground in an old iron mine to provide a shield against cosmic rays and is kept at a temperature of a fraction of a degree above absolute zero, which is much colder than outer space. If WIMPS are the dark matter particles, they should be detected in laboratory experiments on Earth within the next decade.

The ancient Greeks laid the foundation for scientific discovery and began to question why things happen in our world; they wondered about our place in the grand scheme of things. They invented trigonometry and geometry and used it to understand that our planet was just a small spherical rock orbiting the much larger and very distant Sun. They believed that our Sun was just one star among an immeasurable number of other stars that shine in the night sky. There were few new insights into the universe for almost two thousand years, a period of scientific darkness that need not have happened. It finally ended when great scientists like Descartes, Galileo and Newton began the second period of scientific discovery in our history. They showed how moving objects follow well-defined physical laws and that those laws allow the behaviour of matter to be predicted, through cause and effect, through the force of gravity. I have tried to show in this chapter how this knowledge led to our current understanding of the universe.

As we look out into our universe with our observatories and telescopes, we soon realise that our home is not special at all. There is nothing unique about our Sun, or our location within the galaxy. Our "special star", on which all life on Earth depends, is similar to a multitude of others that surround us, many with their own planetary systems. There is nothing extraordinary about the galaxy in which we live or where our galaxy is located in the universe. We have realised that the universe is very old, that it had a beginning and we have measured its age and size. Today, we are at the culmination of a renaissance that has led to our eventual understanding of our place in time and space. Before we delve into its history, contemplate for a moment our insignificance in this incredibly vast and old universe.

Chapter 3
The Big Bang

B. Moore, *Elephants in Space*, Astronomers' Universe,
DOI 10.1007/978-3-319-05672-2_3,

The intensity of learning advanced physics, mathematics and astrophysics within three years at university eased during the summer vacation. I had decided to cycle with my girlfriend Jayne, from Newcastle in North East England, to the Dordogne in France, passing through Germany and Switzerland on the way. I wanted to see the famous prehistoric cave paintings at Lascaux. The journey down the spectacular, castle-adorned Rhine valley was as memorable as the sweet wine "mit Prädikat". It was a cloudy day when we finally emerged from the Black Forest and encountered the foothills of Switzerland, which are higher than the tallest mountains in England. Reaching Interlaken was hard work, but the scenery was beautiful as we cycled along quiet roads winding through fields as neat as gardens, passing through villages lined with fairy tale Swiss houses. The road began to climb steadily up, following the valley carved by the turquoise-coloured water that has been flowing from the high glaciers for the past hundred thousand years. I began to dislike gravity for its relentless pull downwards on our heavily laden cycles. As we slowly climbed higher, the clouds and mist began to disperse and a bright vivid blue sky slowly emerged. My eyes were drawn towards a new sight, jagged snow-covered peaks shimmering in the sun—these were the first real mountains I had ever seen, and their sheer scale and size left me in awe. It was the beginning of an obsession. As I write these words today in the village of Grindelwald, Switzerland, I gaze out of a room with a view of the very same mountains, across to the steep rocky buttress of the north face of the Eiger, a tireless landscape whose shades and colours change by the hour. All those years ago, I had no idea that my seemingly random path through life would bring me back here as a professor of astrophysics in the same institute at the University of Zurich where Albert Einstein and Erwin Schrödinger worked years ago.

At the instant in time that we call the beginning of the universe, space, time and matter appeared and did so with quite a spectacular entrance. We do not know why or how the universe came into existence. We do not know why it has the properties it has or why it has the conditions for life as we know it to be able to develop and evolve. We do not know what (if anything) existed before our universe. It might be the case that these questions may not be answered at a satisfactory level for another century or perhaps for another two thousand years. Perhaps they will never be answered. We may develop and propose theories, but it may be difficult, if not impossible, to verify them via observation or experiment since there may be no remaining detectable signatures from the moment that time began. It is even less likely that we can ever test our theories of what happened before the existence of time. Perhaps the universe just suddenly appeared without a reason. Many scientists are happy to accept that the quantum mechanical view of the world does not necessarily require "cause and effect". In the quantum mechanical world things can happen spontaneously and randomly, or at least they appear to do so. I think that would be a very unsatisfactory description of how things behave, but it may be one with which we have to live. I will have more to say about quantum mechanics and determinism later in this book.

Even though I lay bare the fundamental questions that we cannot (as yet) answer, it is a remarkable achievement of humankind that we understand as much as we do about the universe. We are fairly confident that we can reconstruct its history 13.8 billion years back in time, to within a millionth of a second after it all began. In fact, so much happens within the first few minutes of "the beginning" that many books and tens of thousands of research papers have been written about the fascinating processes that took place during this short interval of time such a long time ago.

In the next two chapters I would like to take you on a rapid journey through the entire history of the universe. It is not an easy journey to make, or an easy one to write. Condensing the findings of generations of scientists is quite a challenge. The ride is tough but spectacular. So hold on tight!

It is difficult to imagine the conditions that existed in the early universe. There is no analogy I can make with anything that you or I have experienced. Just one second after the big bang, our entire visible universe was only a few hundred thousand kilometres across and could have fit within the planet Jupiter. All the material that eventually formed our entire galaxy occupied a volume no larger than a cup of coffee. It is difficult for us to imagine squeezing a single house into a coffee cup, let alone our galaxy with its hundred billion stars. Now you can at least imagine that the conditions in the early universe were not at all like those of our surroundings at present. The densities and energies were so high that interactions took place between particles that could not occur today.

It is quite strange to realise that atoms are mainly empty space and that their constituent particles can be squeezed closer and closer together given enough energy to overcome the strange forces and rules that act to hold them apart. As particles are forced closer and closer together, different reactions can take place, exotic new structures can form and strange new phenomena appear. It is not only in the early universe that matter encounters these extreme conditions. As stars evolve and die, their central cores can collapse into an extremely dense structure such as a neutron star. A teaspoon of a neutron star weighs as much as ten million tonnes. That is equivalent to squeezing the entire human population into a space the size of a single sugar cube!

This is not theory or speculation. We have observed and characterised these fascinating objects throughout our galaxy, and we can completely understand all of their properties using the level of physics taught at university. Even more extreme are the supermassive black holes that astronomers have found at the centres of galaxies, some containing a mass equivalent to a billion stars. The force of gravity may be weak over large scales, but it can be intense when enough matter is present. It can overcome the strong forces that hold atomic nuclei together. When this occurs, there are no other forces that can prevent the matter from continuing to collapse. . .until eventually it reaches a singularity—a black hole.

The big bang model of the universe provides a framework for us to understand its history. As the universe emerged from its dense, hot, fireball-like state and expanded to its present size, it left behind signatures and clues to its past. We study the conditions in the early universe using our knowledge of mathematics and physics, and we use supercomputer simulations to compare our theories with all the observational evidence we find. We can write down equations and formulas that describe how matter behaves, and extrapolate backwards in time to calculate how the universe evolved to its present state. We use sophisticated telescopes on mountain tops and satellites in space to observe how the universe has evolved over time. We try to recreate the conditions present in the early universe using particle accelerators to smash atoms and particles together, to test our theories for the origin and nature of matter.

With these tools we can predict the sequence of events and reconstruct our cosmic history. Like detectives, we seek clues and evidence for our models and theories. As we take our journey through the history of the universe, I will describe the main eras and phenomena that occurred. I will explain further evidence that supports the big bang model, and you can be the jury. Unfortunately, I cannot offer an alternative scenario to explain the observational data that we have acquired—there are no competent alternative scientific ideas to the big bang. We are now moving forward into the new millennium with fresh challenges, to try to understand those unsolved questions that I outlined above, along with other questions about our universe that we will might be able answer in the future. For example, Why is there more matter than antimatter in the universe? Why do particles have mass? Why do the constants of nature have the values they do? What is the nature of dark matter and dark energy? Answers to any of these questions would surely be deserving of a Nobel Prize.

The big bang is not a proven "fact" because it has yet to be described by a complete theory. Nonetheless, based on our current knowledge and scientific methods, it is an extremely successful model for describing how the universe behaved just after it came into existence until today. As the structure of the universe arose through the interaction of particles driven by the force of gravity, so life also developed on Earth driven initially by the electromagnetic force which enabled the first complex replicating molecules to form. Evolution via natural selection is also a model. It is supported by all the evidence that we have gathered and provides an excellent description of how life developed on Earth, just as the big bang provides a framework for understanding the origin and properties of the universe. Evolution also does not yet provide a complete picture of how life appeared; some important steps that need to be understood, especially about its very early origins, are still missing.

No single observation suggests that the big bang model is not a correct description of how the universe has evolved. This does not mean that we should not probe deeper or ask harder questions about our origins. As scientists we should question everything and never be satisfied until enough evidence is accumulated in support of a theory that it may then be regarded as fact. And even then, it is wise to keep an open mind. Newton gave us a simple law of gravity that describes how matter attracts matter under most conditions. This held for several hundred years, even though it actually represents an empirical law proposed by Newton to explain the orbits of the planets. At the beginning of the twentieth century Newtonian gravity was replaced with general relativity, which offers a much more precise theory of gravity that incorporates the effects of matter on the geometry of space and on the passage of time. Einstein's remarkable work on general relativity contains all of the successes of Newton's ideas together with a lot more.

Taking general relativity into account is essential even for practical purposes. A good example is the global position system (GPS) used for navigation, surveying and accurate timekeeping. It works via a network of satellites in orbit around the Earth, each transmitting its location and an accurate time signal using an onboard atomic clock. The signal is decoded, allowing a person's position in three-dimensional space to be determined to an accuracy of just a few metres. Since the satellites are moving at 14,000 kilometres per hour relative to the observer at rest on Earth, they suffer from measurable time dilation—the fact that time and space are closely linked together according to Einstein's theory of special relativity. This causes the GPS satellites to lose about 7 microseconds a day. The clocks are also affected by the Earth's gravitational field, which slows down time the closer you are to its surface. This results in the GPS clocks speeding up by 46 microseconds a day relative to our clocks on Earth. Both of these effects are measured and corrected for by the GPS system.

At some point in the future, general relativity will be incorporated into a new theory of gravity that will also include the effects of quantum mechanics to create a unified theory of quantum gravity. But whatever theory emerges, it must include all of the successes of Einstein's work as well as those of Newton's. As we slowly make progress in understanding how and why the universe came into existence, we may need to modify our big bang model and make a refined theory that more accurately describes our universe. Nonetheless, all the salient points of the big bang picture must be preserved within whatever refinements are made.

One modification that is being incorporated into the big bang model by scientists today is the effect of Einstein's cosmological constant, or dark energy. The rate at which the universe was expanding slowly decreased over the first eight billion years or so as a consequence of gravity, which relentlessly tries to attract all matter together. But around five billion years ago it appears that the expansion mysteriously started to speed up. The universe is increasing in size, and the rate at which it is increasing is getting faster. I will have a lot more to say about that in the final chapters since it impacts on the future of our universe.

This brings up a conceptual problem about the big bang that often confuses people, and for good reason. It is nothing like a giant explosion at some central point from which matter is flung apart into an existing space. The big bang did not happen at one point; it happened at every point! Rather, it is space itself that is being created in between all the matter in the universe. The stretching of space in the universe is often likened to dots drawn on the surface of a balloon that is being inflated. Viewed from any point on the surface, all the other dots appear to be moving away as the balloon's surface is stretched and increases in size. But with this analogy we have to ignore the centre of the balloon. There is no centre to the universe from which all points are expanding; it is the surface of the balloon that gives a two-dimensional analogy to our three-dimensional universe. A better model would be to liken the expansion with an infinitely large cherry-filled fruitcake that is expanding as it cooks in an infinitely large oven. No cherry can be identified as being at the centre of the cake, each one would observe all the others moving away from it, and more distant cherries would be seen to be moving away faster.

So from wherever an observer were to stand in the universe, they would see distant galaxies moving away from them, carried away by the expansion of space. They would also see that more distant galaxies appear to be moving faster away, as expected: something ten times as distant has ten times as much space in between it and us, so it appears to be moving away ten times as fast. As we compare relative velocities across larger distances, we eventually come to a distance beyond which those distant objects will have velocities faster than the speed of light. This paradox appears to violate the fundamental principles behind Einstein's theories of special and general relativity that nothing can move faster than light. We can resolve this dilemma once we study Einstein's principle a little more carefully. What we should say is that objects cannot be accelerated through existing space to a speed faster than the speed of light. Or we could rephrase the statement to say that information cannot be exchanged at speeds faster than the speed of light. The furthest we can see out into the universe is the distance that light has travelled in 13.8 billion years. We have no way of knowing if the universe is ten times larger or infinitely large.

While this all makes sense if we accept that it is "space" that is expanding, it is very frustrating to me that, at a deep level, we really do not understand what space is. Neither time nor space existed before the big bang. I cannot explain to you precisely how it is stretching and what exactly it is that is being stretched. I could tell you how the surface of a balloon stretches due to the uncoiling of rubber molecules held together by intermolecular forces. But that is nothing like empty space, which we can characterise mathematically and geometrically, yet still not compare to anything physical with which we are familiar.

We can conveniently separate our cosmic history into time intervals during which different physical processes were important. In this chapter I want to review what we know about how the universe evolved, starting right back when it came into existence and continuing over the first few hundred thousand years. This is the era in which atoms were made, yet the universe was smooth with no structure. In the following chapter I will describe the remaining history of the universe to the present epoch, the time during which stars, planets and galaxies formed. Despite there being many good popular books on these topics, cosmology is a rapidly advancing research area. I want to spend more time on results that have emerged over the past decade that have changed our understanding of the history of the universe and led to an understanding of how the universe will evolve in the future.

The reason that different physical processes occur at different times over the history of the universe is that the temperature of particles and photons decreases as space expands. Lower temperatures and greater volume mean that particles collide with less energy and less frequently. The average temperature of the universe today is 2.7 Kelvin (−270.3 degrees centigrade). That is the temperature you would find if you put a thermometer in empty space well away from a nearby star. I will explain how we know this later. If you could put the same thermometer at the same point in space in the early universe, say one second after the big bang, you would measure a temperature of 10 billion Kelvin!

Temperature is a measure of the kinetic energies of particles. When the air around us is at a higher temperature, we perceive warmth. This is because the air molecules are constantly colliding with us trillions of times a second and transferring their kinetic energy (that is the energy due to the fact that they are moving) to the epidermal layer on the surface of our skin. Since energy is always conserved, when a moving particle stops, it transfers its energy to whatever stopped it from moving. This "feels" warm to us since our nervous system can monitor temperature: the warmer surroundings of the nerve endings trigger electrochemical impulses that travel along our nerves to the neural network in our brain. We evolved with the ability to sense temperature: it is useful to know when things get too hot or too cold so we can prevent damaging our bodies through burning or freezing.

To see how the temperature of the universe decreases as it ages, consider what is happening in a small region of space. As the universe grows in size, or more correctly, as the amount of space between particles of matter increases, their relative speeds decrease and fewer collisions take place between the particles. Consequently, the average temperature and density of the universe decrease with time. Photons also lose energy as the universe expands and at a rate that is faster than that of the particles of matter. This is because their wavelengths stretch along with the space and their energy decreases. It is as if photons are somehow pinned to space and as space expands, the photons transfer their energy into the newly created space. The energy of a photon is proportional to one over its wavelength. This means that if the universe were to double in size, the wavelengths of any moving photons would be twice as long and their energy would be halved.

Using the equations of Friedmann and Lemaître (which tell us how the size of the universe changes with time), we can calculate physical conditions such as density or temperature at any epoch in the history of the universe. We can extrapolate backwards billions of years to a time that is just a tiny fraction of a second. But as we get closer and closer to "the beginning", our understanding becomes less and less certain. The reason for the lack of a complete description of the big bang is that the conditions become so extreme that we start to find major gaps in our knowledge of fundamental physics. We simply do not have a unified theory of the four forces that we know occur in nature, and we do not understand how quantum mechanics and gravity work together under the most extreme conditions.

While we perceive time as progressing in a linear way, with one second following another, it makes physical sense to dissect the history of the universe using time measured "logarithmically" in powers of 10. This allows us to neatly divide up past epochs into well-defined eras during which different physical processes occur.

The Planck Era: T = 0 to 10^{-43} seconds

This first era of the universe is perhaps the shortest interval that is sensible for consideration with our current understanding of time. This timescale, called the Planck time, is the time taken for a photon to travel one Planck length, which is a natural distance that results from a simple combination of the fundamental constants of nature: the speed of light, the gravitational constant and Planck's constant. Planck's constant is the size of a "quantum" in quantum mechanics that describes the allowed energies that particles and photons can have. During this period we believe that all the forces of nature worked together as a single unified force. It would be a truly remarkable achievement if we could one day understand this first epoch of our history.

The Grand Unification Era: 10^{-43} to 10^{-36} seconds

At this time, just 10^{-43} seconds after the big bang, the temperature of the universe was an incredible 10^{27} Kelvin—a billion billion billion degrees. Particles as we know them were not yet in existence, but we think that the gravitational force became distinct and unique as it decoupled from the electromagnetic force and the strong and weak forces. We know very little about this epoch and absolutely nothing about what occurred before.

The Inflationary Era: 10^{-36} to 10^{-32} seconds

The universe cooled down just enough to allow the strong force to become separate from the weak force. It has been proposed that an instability in the vacuum energy drove an incredibly rapid expansion of space known as inflation. The dimensions of space increased exponentially, by a factor of at least 10^{26} in a tiny interval of time as the "potential energy field" of the vacuum attempted to reach its most stable low-energy state. The enormous energy of the vacuum now populated the universe with a dense relativistic sea of quark and anti-quark pairs and gluons. Quarks are the elementary particles that make up protons and neutrons, and gluons are particles that act as a messenger particle in the exchange of forces between quarks.

Inflation is a hypothesis that explains several global features regarding the universe. It naturally leads to a space-time that is geometrically flat. Imagine the surface of a balloon that is not inflated but wrinkly and irregular. When the balloon is inflated, the surface becomes smoother. Imagine blowing up the balloon until it is a trillion trillion times larger—the surface around any point on the balloon becomes extremely smooth and flat like a plane. Inflation also provides a mechanism for generating tiny fluctuations in the distribution of matter from which all structure in the universe arose. The early exponential expansion of space during the inflationary epoch is an intrinsic part of the big bang model. But although observational evidence supports this idea, it has not been confirmed per se; rather, the theory is so elegant that it seems it must be correct.

The Electroweak Era: 10^{-32} to 10^{-6} seconds

We are now in the realm of the big bang model that is becoming directly testable. At this time the fundamental particles appeared. The temperature of the universe was still too high to allow the quarks to bind together to form protons or neutrons, and space was filled with a hot quark-gluon plasma. This is the era that the Large

Hadron Collider (LHC) at CERN (the European Organization for Nuclear Research), near Geneva, has been designed to probe, by colliding particles with energies that are similar to the interactions that occurred during this epoch.

Recreating the Early Universe on Earth

The quark-gluon plasma is a strange state of matter that has never been experimentally studied, but the objective of the LHC is to recreate the energies and temperatures that existed during this era of the universe. The LHC is one of the most sophisticated scientific experiments ever undertaken by humankind. It has taken over a decade to construct using the combined efforts of tens of thousands of scientists and engineers from over a hundred countries around the world. The experiment consists of a circular 27-kilometre-long tunnel passing under the Swiss and French countryside, lined with superconducting magnets, for accelerating protons to a speed that is just 3 metres per second less than the speed of light (almost 300 million metres per second).

The tunnel itself is about as big as the London Underground, but the particles are actually accelerated by an electrical field in a tube no wider than a fire hose. The beam tube holds an ultrahigh vacuum, as empty as interplanetary space; otherwise the protons would collide with molecules of air. The proton beam itself consists of about 3,000 packets each containing a 100 billion particles: Half of the packets are sent clockwise and the other half counter-clockwise. The actual clouds of particles are about a millimetre across and weigh less than a single blood cell, but at full speed the proton beam has the energy of a 400-tonne train travelling at 200 kilometres per second! Remember, energy is mass. To keep the particles within the ring, there are over nine thousand magnets along the route which focus the particles and eventually concentrate the packets to the width of a human hair before they collide. The superconducting magnets each weigh over 27 tonnes and are cooled down to −271.3 degrees centigrade to increase their efficiency. The magnets need to be extremely powerful since, as the particles move faster, their inertial mass increases 10,000-fold. That's a consequence of Einstein's relativity. Speed, mass, energy and time are interdependent: the faster you travel, the more you weigh and the more your clock slows down.

It would take the Space Shuttle about five seconds to travel a distance the length of the LHC tunnel—it travels ten times faster than a bullet. That is impressive, but after being accelerated for about an hour, the packets of protons reach their top speed: they are travelling so fast that they travel around the 27-kilometre tunnel over 11,000 times in a single second!

Finally, when they cannot be made to move any faster, the two particle beams are made to crash into each other inside one of the giant detectors which occupy an underground cavern as big as a cathedral. When two packets of protons collide, there will only be about 20 collisions between a trillion crossing particles. But since the packets cross 11,000 times per second, hundreds of millions of collisions take place. These collisions between particles occur at the highest energies ever achieved on Earth, and are comparable to those taking place during the first second of the universe. The giant detectors record all of the details of the collisions, generating 15 million gigabytes of information per year. Now you can understand why I describe the LHC as the most sophisticated scientific experiment ever undertaken by humankind!

Protons and neutrons are the particles at the heart of every atom, but they themselves are made of three quarks. Quarks have various intrinsic properties, such as flavour, charge, colour and spin. These names bear little relation to the physical meaning with which we usually associate them, but describe instead their behaviour mathematically and how they interact with each other and with other particles. The collisions that occur between the protons in the LHC probe their internal structure and the nature of the quarks from which they are made. Democritus and the ancient Greeks speculated that matter was made up of tiny particles called atoms. I wonder what Democritus would say if he could see the LHC today and could look inside the very particles that constituted his "atoms".

One of the main goals of the LHC was to detect something called the Higgs boson. Bosons are subatomic particles that are responsible for transmitting the fundamental forces of nature—the electromagnetic, weak and strong forces. They act like messenger particles when normal atomic matter interacts. There are five basic bosonic particles: photons, gluons and the W, Z and Higgs bosons. All of these except for the Higgs boson have been detected in experiments. The Higgs boson is an important particle to study since we think it provides the mechanism by which particles acquired their "masses". From the billions of collisions that occur, the signature of a Higgs boson was expected to be seen only once every few hours. By 2013 enough data had been collected to enable us to decide whether or not it actually exists and the Nobel Prize in Physics was duly awarded.

The Hadron and Lepton Era: 10^{-6} to 10^{1} seconds

One second after the big bang the temperature dropped to a comparatively cool 10 billion Kelvin. The energy in particle–particle collisions was low enough such that trios of quarks combined together to form individual protons and neutrons. Until this time, electrons were combining with anti-electrons to produce neutrinos and anti-neutrinos. The temperature was so high that the reverse reaction was also taking place. Two seconds after the big bang the temperature had decreased just enough to impede this reaction, and the neutrinos were free to travel the universe in all directions and from all places. Today these relic neutrinos fill space—there are trillions of them passing through your bodies every second. Unfortunately (or fortunately for us), it is extremely hard to detect them since they interact very weakly with atomic matter; they could easily travel through a piece of lead that is one light year in length without colliding with a single atom.

The Photon Era: 10^1 to 10^{13} seconds

From 10 seconds to 380,000 years after the big bang, photons ruled the landscape, and their energy drove the early expansion. (In fact, photons outnumber protons and neutrons; for each proton or neutron that exists in the universe, there are ten million photons filling space.) Three minutes after the big bang the temperature and conditions in the universe resembled those that occur at the centre of the Sun. Prior to this point protons and neutrons literally had too much energy to stick together to form atomic nuclei. As the expansion of the universe continued and particle interactions occurred with less energy, there was a brief 17-minute window where nuclear fusion occurred, resulting in the formation of the nuclei of hydrogen, helium and a trace of lithium. At the end of this period, the density of the universe had decreased such that further collisions between nuclei become too rare to create any of the heavier elements in the periodic table.

At this time, the electrons still had too much energy to attach themselves to the nuclei, so they remained diffusely distributed. But they interacted strongly with photons, constantly colliding, transferring momentum and energy, and keeping the gas ionised like a plasma. The narrowness of this window of time during which the basic nuclei formed allows us to make a precise prediction of their relative abundances. According to everything we know about particle physics and cosmology, nearly four times as much hydrogen as helium was created during those brief 17 minutes. This is exactly the ratio that is observed in our universe, providing very strong evidence for the big bang model and the fact that the universe was once dense and hot enough to create atoms.

The Cosmic Microwave Background

As the particles of matter and antimatter collided and annihilated each other, they released energy in the form of photons. We do not know why, but fortunately for us there was a small asymmetry between the amount of matter and antimatter in the early universe—for every billion particles there was about one extra particle of "normal" matter. Just one minute after the big bang no antimatter was left, and the high-energy photons that originated from the final matter–antimatter annihilations filled the universe. After 380,000 years the temperature of the universe had cooled to 3,000 Kelvin. The energies of electrons had decreased sufficiently to allow them to bind to the atomic nuclei, and neutral atoms of hydrogen, helium and lithium formed. But none of the other elements of the periodic table were made until the first stars formed after about half a billion years. Prior to this time the photons were colliding and bumping off the nuclei and electrons. Quite suddenly, the universe became "neutral" and the photons could move freely. From this time onwards the photons no longer had anything with which to interact, and they continued their timeless journeys unimpeded, moving in all directions. As space continued to expand, however, their energies were decreasing.

These are the cosmic microwave background photons whose existence was predicted in 1948 by George Gamow and his students. They were detected accidentally by Arno Penzias and Robert Wilson in 1964. They provide the strongest evidence for our big bang model of the expanding universe. Gamow, a student of Friedmann, was another of the mathematical geniuses who emerged from Russia. His paper titled "The Origin of the Chemical Elements" detailed how the abundance of hydrogen and helium could be explained by reactions between particles that occurred during the first minutes after the origin of the universe when it was filled with high-temperature photons. He suggested that the residual photons or afterglow of the big bang must have permeated all space and would have cooled down to just a few degrees above absolute zero. This is the average temperature of the universe today. Most theorists thought that this radiation would be too weak to detect, and it did not seem to catch the attention of experimentalists at the time.

Penzias and Wilson, of Bell Labs, were constructing a telescope to observe the universe in the radio wavelengths. They were frustrated by a constant background noise from their antenna which seemed to come from every direction in which they aimed the telescope. At first they thought the noise was due to the pigeons that nested inside the horn-shaped antenna. But Canadian cosmologist Jim Peebles pointed out to them that they might have observed the signals of relic photons from the big bang. The cosmic microwave background photons were predicted to have a characteristic spectral distribution that is detectable in the radio and microwave bands from 0.03 to 300 centimetres, and that is exactly what Penzias and Wilson observed. The evidence was so strong in support of the big bang model of the universe that it immediately put paid to competing theories. The most compelling of these was Fred Hoyle's steady-state model, but it could not explain the observed photon background.

It was quite a feat for theoretical cosmologists to speculate the existence of phenomena in our universe that would not be observed until several decades after their work. In the 1970s the theoretical cosmologists Peebles and Yakov Zel'dovich realised that entire galaxies could be formed by the attractive force of gravity acting to pull matter together from a large enough region of the universe. All that was needed were some small irregularities in the early distribution of matter to act as seeds for the subsequent growth of structure. This is an elegant idea for how galaxies form and one that we have recently verified using supercomputer simulations. As the photons travel across these regions of matter irregularities, they gain or lose a small amount of energy due to changes in the curvature of space caused by gravity. The small fluctuations in the distribution of matter lead to small fluctuations in the temperatures of the microwave background photons. In 1983 a Russian satellite called RELIKT-1 was launched in the first attempt to measure them. The researchers involved in the effort came tantalisingly close and even claimed to detect the temperature fluctuations, although it was not until 1992 that NASA's COBE (Cosmic Background Explorer) satellite achieved an unambiguous detection. After the discovery was announced Stephen Hawking said in a statement, "It is the discovery of the century, if not of all time".

The importance of the cosmic microwave background in verifying the big bang cannot be understated. The WMAP (Wilkinson Microwave Anisotropy Probe) space satellite, launched in 2001 to observe the universe in microwaves, gave a detailed picture of the conditions that were present just a few hundred thousand years after the big bang and has provided precise information about the initial conditions from which all structure in the universe was formed. There is a new satellite mapping its structure in exquisite detail right now, the European Space Agency's Planck mission, launched in 2009. Why the continued interest? The Planck satellite was designed to extract all the information possible from the microwave background, and to measure the cosmological parameters with extreme precision. It has probed a range of astrophysical phenomena, including theories about the very early universe, the origin of cosmic structure, the cosmic inventory (including dark matter and dark energy) and perhaps the detection of the first stars to shine in the universe.

During the first 380,000 years since the beginning of time, a complex sequence of interactions between individual elementary particles took place that led to the formation of atomic matter. The universe expanded and cooled to a size such that the particle–particle interactions that led to new forms of matter ceased—the composition of the universe was fixed. At this time there were no complex structures, no stars, planets or galaxies. There was no oxygen, carbon or any of the atoms in the periodic table apart from lithium—there was no life.

The individual atoms of hydrogen, helium and lithium, together with the dark matter particles, were all smoothly spread out through space. But the universe was young, and there was plenty of time for structure to emerge as matter interacted via the gravitational and electromagnetic forces. This era is known by astronomers as the "dark ages", which is quite different from the term I used in the first chapter to describe the period of intellectual darkness in human history. In astronomical terms the dark ages denote the period following recombination, the time when the microwave background emerged and the atomic material became neutral. It lasted for a full 500 million years prior to the formation of the first stars, and during this period the universe was quite literally, dark. In the next chapter I will review how stars, planets and galaxies developed from this very smooth and dark universe.

Chapter 4
The Origin of Stars, Planets and Galaxies

B. Moore, *Elephants in Space*, Astronomers' Universe,
DOI 10.1007/978-3-319-05672-2_4, © Springer International Publishing Switzerland 2014

I surprised myself by achieving a first-class degree in physics and astrophysics and briefly debated between joining British Aerospace to design space rockets and the alternative of doing a PhD in cosmology to hopefully unlock some of the secrets of the universe. I hesitated only for a moment before beginning my second steep learning curve: figuring out how to do original research. During my PhD studies I was sent to a summer school in Les Houches to learn from famous scientists, but the view from the windows of the lecture hall was more than just a little distracting. Mont Blanc towered above the surrounding peaks, so close but so far away. During the coffee breaks a small group of students would congregate outside and stare at up at the imposing mountain. "Let's climb it," I suggested, only half joking, but the others felt the same desire. We were young and foolhardy, never having climbed a mountain before. But one of the students at the school was an experienced mountaineer from Switzerland who agreed to be our guide. The following weekend we began the long traverse of Mont Blanc via the Goûter ridge and descending via the Aiguille du Midi. The climb was exhausting, another struggle against the relentless force of gravity; but it was breathtakingly beautiful. We reached our first camp by sunset and had only a few hours of restless sleep before we were woken at 2 am to begin the steep ascent to the summit. Roped together and armed with crampons and ice axes, we followed the line of headlamps shining along with the moonlight in the dark, from even earlier rising climbers. The surrounding mountains glowed dimly from the bright starlit sky, and for the first time I saw clearly the spectacular disk of stars of our Milky Way. I felt insignificant and wondered why we seemed so alone in our galaxy. I will never forget the narrow ridge that led to the icy summit, with sheer drops down both sides and the spectacular panoramic views of the surrounding jagged peaks rising up through an ocean of clouds.

Our bodies are made up from about 10^{28} different molecules which are themselves composed of individual atoms consisting by weight of 65 percent oxygen, 19 percent carbon, 10 percent hydrogen, 3 percent nitrogen and 3 percent everything else. Over half of our body weight is made up of water molecules (H_2O). Since hydrogen is the lightest element, it does not contribute very much to the total body mass of a human being. But if the number of atoms are counted, then 60 percent of the particles in our bodies are actually hydrogen. The hydrogen atoms in the water were made just a few minutes after the beginning of the universe. Thus a large fraction of your body is 13.8 billion years old! The oxygen and the rest of the atoms in our bodies were forged in stars and not just one star—they originated from literally thousands of different stars from all over the galaxy. By mass, 90 percent of your body is made of stardust, elements created in the nuclear furnaces at the centres of stars that lived and died between 4.5 and 13.8 billion years ago.

When stars die, they recycle some of their gaseous material into the galaxy by violently blowing away their outer layers. Stars that are forming today are assembling from a mixture of primordial hydrogen and helium, elements created in the big bang, together with heavier atoms that were forged at the centres of other long-dead stars. The elements heavier than hydrogen and helium in our Sun, in the planets and our bodies, were made within earlier generations of massive stars that burned brightly and died quickly. At the end of their lives these stars literally exploded as spectacular supernovae and scattered the elements they made throughout the galaxy. The particles in your body have come together over billions of years on very different paths, traversing the galaxy and ending inside our solar system and ultimately coming together to form your conscious self. This is truly mind-blowing stuff!

For over four billion years our Sun has provided energy for our planet. As the Sun burns through its atomic energy resources, it is slowly getting hotter. The conditions on Earth will gradually become harsher and make it more difficult for life to exist on its surface. In about one billion years the temperature on the Earth's surface will exceed 100 degrees centigrade and the oceans and lakes will begin to boil away. In a further seven billion years the Sun will stop shining. As the stars in our galaxy fade and fewer new stars form, the outlook for life to continue to thrive looks bleak. In this chapter I would like to explain how our Sun and solar system formed and how stars can shine for as long as they do.

The Emergence of Structure in Our Universe: The Final 13.799620 Billion Years

We know from cosmic microwave background fluctuations that very early in the universe, matter was not distributed perfectly uniformly. Variations in the spatial distribution of particles are thought to have arisen as a consequence of quantum effects during the Planck epoch—rather like random noise—an imprint of the first instance of time in our universe. The exponential expansion of space during the following inflationary era would literally stretch these microscopic quantum fluctuations to macroscopic scales, leading to some regions of space with more particles on average and other regions with less. In some places the amount of matter is large enough so that it can locally reverse the expansion of the universe, pulling together material from the surrounding region. Within a billion years of the big bang, gravity began to assemble the first complex and massive structures in our galaxy—the dark matter and atomic particles collapsed into giant structures that are thousands of light years across. Gas accumulated at the centres of these structures and formed stars and eventually galaxies.

In the last decade we observed and characterised these initial conditions from which galaxies formed. We have also quantified the composition of the universe in terms of its matter and energy densities. We have all the information we need to determine how the universe evolved over time to the present day and to enable us to calculate how stars and galaxies form. In principle, it is just a matter of working out how the particles of matter moved as they interacted with each other via the gravitational and electromagnetic forces. In practise though, the equations are far too complex to solve by hand. Consequently, we rely on sophisticated numerical codes running on the world's most powerful supercomputers to calculate how stars and galaxies form within our expanding universe.

Our calculations are made within the "standard cosmological model", which is sometimes called the ΛCDM model, where Λ (the Greek letter *lambda*) is the dark energy and CDM is cold dark matter, the simplest form of matter that could make up the missing mass in clusters and galaxies. Our standard ΛCDM model for structure formation provides a scenario within which to test our ideas of galaxy and star formation. It has already achieved a huge amount of success, from matching the topology of the large-scale distribution of galaxies to reproducing the formation of individual galaxies like our own Milky Way.[1]

Turbulence and Star Formation

If you look carefully at the stars in the night sky, you will see that they have many different colours. As stars use up their fuel and evolve over time, their temperatures and chemical composition change. We see this visually in the spectrum of their light, which we perceive as a colour. What you are seeing is a mixture of stars of different ages and masses; some stars are being born today and others are over ten billion years old. Right now, stars are forming in our galaxy at a rate of about one per year. That does not sound like a particularly large number, but over several billion years it represents a lot of stars.

The stars in our galaxy are forming within what we call the diffuse interstellar medium. This is a huge gaseous layer of a mixture of primordial hydrogen and helium together with stardust that circles our galaxy, filling the space between the stars. Its density varies from just a few hundred to several million atoms per cubic centimetre, far more tenuous than the air we breathe. But our galaxy is large, and there is at least enough atomic material to form another ten billion stars similar to our Sun.

The gaseous interstellar medium is turbulent—a physical process that thoroughly mixes the atomic material and enables new stars to form. Turbulence is what makes aeroplanes bounce around under adverse weather conditions. It mixes smoke with air or makes the long lasting wake behind a boat's propeller. It is also the mechanism that mixes cream with coffee when it is stirred with a spoon. If cream were poured slowly into a cup of coffee, it would just sit there at the bottom of the cup. After a few stirs of the spoon though, the cream becomes completely and uniformly mixed into the coffee. That is quite something: The spoon is a solid object only a couple of centimetres across, yet in a mere one or two seconds it thoroughly mixes 10^{25} individual molecules all the way down to the mean spacing between the molecules in the cup (about 10^{-10} metres). This mixing is achieved through turbulence: The spoon causes the liquid in the cup to spin and rotate. The swirling motions start out on the largest scale (that of the entire cup) and then propagate to smaller and smaller scales via a cascade of vortices. The result is a vast number of ever smaller whirlpools that stir and mix the liquid right down to the microscopic molecular level. This must be the real reason that James Bond likes his vodka martini shaken and not stirred: He doesn't want a complete mix!

[1] Movies from our computer simulations which show the formation of dark matter haloes and galaxies can be seen on the author's homepage: http://www.astroparticle.net

Turbulence is a very complicated but important phenomenon to understand, yet we only have a qualitative idea of how it works. We know the exact equation that the gases and liquids must obey (called the Navier–Stokes equation), but nobody has been able to find solutions to it. It is an important topic not only because it is the reason that stars form but also because it affects many aspects of our lives, from the design of aircraft and motor cars to the understanding of the weather and hurricanes. The Clay Mathematics Institute in Cambridge, Massachusetts, offers a million-dollar prize, not to someone who can mathematically solve turbulence but for the first person to prove that solutions actually exist!

The turbulent motions of the swirling gaseous material in our galaxy are driven by the powerful blast waves from exploding stars and the gravitational forces that act between them. These are the cosmic coffee spoons that stir the interstellar medium. The result is a cascade of enormous vortices that mixes the interstellar medium. The elements of which you are made originated from numerous different exploding stars from across the galaxy. Some of these vortices accumulate enough gas so that they become gravitationally unstable—these dense clumps of gas are massive enough to collapse under their own weight.

As the gas collapses it heats up because more matter being compressed into a smaller volume causes the gravitational force to increase and the particles to fall faster and faster together. Turbulence converts the infalling flow of the gas into random motions of the particles, raising its temperature. For the gas to continue to collapse to smaller and smaller scales, it has to cool and dissipate its energy. It does this by radiating photons. When a star is forming, over 10^{57} atoms collapse together; their density increases from a thousand particles per cubic centimetre to about one gram per cubic centimetre—an increase in density of a factor of 10^{20}! At the centre of the proto-star the pressure steadily increases because of the continuous pile-up of material squeezing and compressing the gas. Its temperature soon reaches thousands of degrees, causing it to glow in visible light. A giant turbulent gas cloud may contain many hundreds or thousands of these collapsing proto-stars. You may have seen photographs of such regions taken by the Hubble Space Telescope. My own favourite image is that of the Orion nebula. The clouds of dusty purple-red-brown gas illuminated by the newly forming stars are a work of abstract art in its purest form. In my opinion nothing as wonderful as this has ever been painted by a human artist.

Fusion Reactors

With gravity relentlessly pulling matter inwards and being compressed from a trillion trillion tonnes of infalling material, the central proto-star becomes denser and hotter until after a few million years it reaches a temperature of 10 million degrees Kelvin. At this time a new energy source switches on: nuclear fusion, the conversion of matter into pure energy. These temperatures are only achieved if the collapsing core mass is larger than about one tenth of the mass of our Sun. Proto-stars with a mass less than this cannot ignite the fusion process. They are called brown dwarfs. They simply fade out of sight, slowly radiating away the energy they have acquired during the gravitational collapse. There are at least as many of these "failed stars" as there are stars in our galaxy.

The Sun is powered by the energy released when two nuclei of hydrogen are forced close enough so that they combine as one, fusing together to form a single nucleus of helium. The sequence starts out with two protons that smash into each other with sufficient energy that one of them undergoes a transmutation into a neutron by emitting energy in the form of a neutrino. This proton–neutron pair is an isotope of hydrogen called deuterium, which itself will fuse together with another deuterium atom, thus creating helium. The mass of the resulting helium atom does not weigh as much as the four protons at the start of the process. It is a tiny bit lighter, about one percent less than at the start. That one percent of mass was lost when the strong force was overcome during the coalition of the two separate nuclei. This small amount of mass is converted in to a large amount of pure energy: neutrinos and photons that provide the heat source emanating from the core of the star. The energy that is released from fusion prevents the star from collapsing any further, and the entire massive ball of hot hydrogen and helium plasma settles down into a self-regulated equilibrium state.

These reactions between particles are very unlikely unless the conditions are quite extreme, for example at the centre of a star. Higher temperatures lead to faster-moving particles that collide more often and with more energy. You might wonder why the Sun does not just explode in a matter of seconds during a chain reaction of fusion as the temperatures get hotter and hotter, creating a massive cosmic fusion bomb. This is because stars regulate themselves via a feedback loop. If the temperature and pressure increase at their centre, the material gets pushed outwards and they start to expand. Expansion leads to cooling, just like the cooling that occurs as the universe expands. This then decreases the number of fusion reactions, and the star contracts again.

Our Sun is made up of 75 percent hydrogen, 23 percent helium and 2 percent of all the heavier atoms of the periodic table, such as oxygen, carbon, iron, gold and uranium. The region that undergoes fusion extends to about a quarter of the Sun's radius where the pressure is so high that the density of the predominantly hydrogen and helium gas is over a hundred times that of liquid water. Every second 600 million tonnes of hydrogen are fused together, of which about four million tonnes are converted directly into energy. That energy is mostly in the form of photons of light in the gamma-ray wavelengths.

It takes a long time for the energy created at the centre of the Sun to reach its surface. The photons travel only a few millimetres before they collide with another proton which scatters the light into a different direction. The photons make what is called a random walk, bouncing this way and that until sooner or later they end up at the surface of the star. Eventually they escape into space. This is a journey that takes up to one million years! The photons from the fusion process are high-energy gamma rays. But by the time they reach the surface of the Sun, they have transferred most of their energy to the protons. This keeps the interior of the Sun extremely hot.

When they finally leave its surface, the photons emerge in the lower-energy optical wavelengths.

Eight minutes later the photons reach the surface of our Earth. The light you see from our Sun today originated at the centre of our star at a time before *Homo sapiens* appeared, when our distant ancestors were beginning to use fire. It is not a coincidence that our eyes are sensitive to the part of the electromagnetic spectrum that is associated with colour: the optical wavelengths. The reason is simply that our eyes have evolved to be the most sensitive to the light which illuminates our planet—sunlight.

The Earth's surface captures less than a billionth of the light coming from the Sun, but that is still a vast amount of energy. If we could capture all of the sunlight that strikes the Earth's surface in a single hour, we could provide the global power requirements of the Earth for one year. In fact, the total energy output of the Sun could support 20 trillion planets like our Earth for 10 billion years. That is an impressive amount of energy. However, our star, like all stars, will eventually use up all of its fuel and will then stop shining. Detailed calculations show that our Sun will die in the year 7,600,000,000 AD. Long live the Sun!

If the Sun suddenly stopped shining, within just a few days the Earth would rapidly cool down to below freezing everywhere on its surface. There would be no climate, no weather, no wind and no running water. The impact would travel rapidly through the food chain: Photosynthesising plants would quickly die, the animals dependent on plants for their food or habitat would all die, the animals dependent on other living creatures for food would all die. The oceans would freeze over, but underneath the insulating ice layer, the water would remain liquid for a long time. Our Earth would resemble Enceladus, the frozen satellite of Saturn. The oceans would not freeze solid for millions of years thanks to the heat energy inside the Earth, which comes mainly from the radioactive decay of elements within it. The hydrothermal vents deep under our oceans could still provide the energy and nutrients to maintain life if it could exist without the oxygen-rich atmosphere on the Earth's surface.

However, there is no danger of this happening for another several billion years since our Sun is less than halfway through its life. Stars live and die by well-understood physical principles. Moreover, we understand in great detail how they shine; they are much simpler to understand than how a simple blade of grass lives and dies. Stars obey the laws of physics that scientists have uncovered in the past hundred years, and we can write down relatively simple equations to describe how they work. These are equations that we can accurately solve to predict how the Sun has changed in the past and how it will evolve in the future.

For most stars, we can only measure mass, chemical composition, surface temperature and luminosity. That is sufficient information for us to ascertain their age and their life span. We can approximate a star by a spherical ball of ionised gas or plasma. It is not a uniform sphere though; its density, temperature and chemical composition vary with radius and over time in such a way as to maintain its equilibrium structure. It is a natural balance between gravity (which always pulls material inwards) and the energy from fusion that prevents the star from collapsing into a much denser configuration such as a black hole. The high-energy photons released during fusion create a repulsive pressure by interacting with the atomic matter, increasing their random motions and their temperature which in turn pushes material outwards.

Initially, the material in the star is well mixed, but its chemical composition slowly changes as hydrogen is converted to helium at its centre. We can write down equations that relate how the temperature and pressure inside the star change with time as nuclear fusion slowly changes its chemical composition, and as mass is slowly depleted and turned into energy and new elements form. This allows us to calculate the "observable quantities" for a star at any point in its lifetime, its energy output and the temperature of its surface. The density, temperature and pressure all increase towards the centre: the temperature of the surface of our Sun is about 6,000 Kelvin, whereas the matter at the centre is at 15 million Kelvin. We cannot put a thermometer at the centre of the Sun. But we can calculate its central temperature because that is how hot it needs to be for fusion to occur and to generate the observed temperature and luminosity at the surface.

The evolution throughout the life of a star is so slow that we cannot measure how our Sun changes over our short lifetimes. We can, however, observe millions of nearby stars that are at different stages in their evolution. The models can then be compared to, and refined by, observations of many stars of different masses and at different stages in their life cycles. For one star at least, we can test the predictions of our models in great detail, and everything seems to agree exactly. Through the remarkable observations of helioseismology we can measure the internal structure and composition of the Sun quite accurately.

Seismology is the technique used to study the internal structure of the Earth using sound waves that pass through its interior. The deepest hole we have ever dug from the surface of the Earth is just a few kilometres deep, so how do we know for certain that the Earth has a solid iron core? The sound waves generated by earthquakes or nuclear explosions travel right through our planet, reflecting off various materials in different ways. The patterns and speeds of the reflected sound waves can be used to reconstruct the internal structure of the Earth. The nuclear blast waves at the centre of a star also cause sound waves that reach its surface. Using Doppler imaging techniques, we can observe the surface of the Sun vibrating and can measure its harmonics and tones, just like the vibrations of a violin string. The surface patterns on the Sun can be used to infer its structure deep within its interior. The data are in excellent agreement with our models and theoretical predictions. In fact, we know more about the centre of our Sun than we do about the centre of our Earth.

Origin of the Planets

Not all of the collapsing gas ends up inside the star. Some of the material remains swirling and spinning around the dense hot core. It is from this material that the planets form. Rotating structures are common in our universe. The planets rotate in the same plane around our Sun, and the stars rotate in a giant disclike structure around our galaxy. The reason for such ordered and fast rotation is one of those conservation laws that we have already come across. I have already described how a spinning object stays spinning forever because of the conservation of rotational motion. The gas cloud that forms a star starts out collapsing from a much larger region, and it has a small amount of rotation resulting from turbulent motions. Since the cloud is becoming smaller and smaller as it cools and contracts, it has to spin faster and faster to preserve the same amount of rotational energy. Ice dancers speed up their spinning motion on the ice in exactly the same way: Starting to spin with arms extended, they slowly pull their arms closer to their bodies which makes them spin faster and faster. In the same way, as the turbulent gas cloud collapses, the surrounding proto-planetary disk ends up rotating about the newly forming star. Spinning material that is loosely held together, like a liquid or a gas, naturally arranges itself into a flattened rotating structure, unlike the ice skaters, who maintain their shape since the atoms in their bodies are held together with strong intermolecular forces. It is visually similar to the way a ball of pizza dough forms a flat disk when it is thrown spinning up into the air.

By the time the star ignites and fusion switches on, the first stage of planet formation is already well under way. The density in the proto-planetary disk is now about one billionth of a gram per cubic centimetre—that is a billion times less dense than the air that you are breathing. Air molecules move about one millionth of a centimetre before they collide with another molecule. This is the process by which a sound wave propagates through the air, and it is set by the speed at which the molecules bump into each other. In just one second after you speak, an enormous sequence of collisions has taken place that has propagated about 340 metres, which is the speed of sound in air and the speed at which the molecules are randomly bumping around. In one second a molecule of air has collided with over a billion other molecules!

The gaseous proto-planetary disk contains almost all of the atomic elements in the periodic table, which combine together to create a large number of complex molecules. The disk is far less dense than air, so its molecules move about one centimetre before they collide with another. It seems remarkable that the forces of nature can conspire to create entire planets from such a diffuse gaseous medium. But over the course of one orbit at the Earth's distance from the Sun, trillions of collisions between molecules will have taken place, during which time molecules and small aggregates of solid material start to form, rather like specks of dust. It is this stage of planet formation that is least understood—the problem being that we can't directly observe this stage of evolution of planets around nearby stars and our supercomputers are not yet powerful enough to fully test this era.

The temperature in an early proto-planetary disk is hot—several thousand Kelvin at the Earth's distance from the Sun. Only rocky metallic compounds, such as iron oxides and silicates, could form in the inner solar system. Icy water-based structures and carbon-based compounds could only condense at distances beyond four times the distance of the Earth from the Sun. Any closer to the newly forming star and they would melt and dissociate and remain in a gaseous form. This is interesting: Carbon forms more compounds than any other atom. There are over ten million known organic compounds, and carbon is essential for life as we know it. Our Earth's rocky crust is almost 50 percent oxygen, which is mainly contained within iron and magnesium oxides, whereas its carbon content is only 0.05 percent. Oxygen is a thousand times more abundant than carbon in the Earth, but in the gaseous interstellar medium from which the planets form, only twice as much oxygen as carbon exists. We might therefore expect that 25 percent of the Earth would be carbon, based on its universal abundance relative to oxygen. In fact, it is just a tiny fraction of the Earth's mass. That is because carbon-based compounds cannot form at the Earth's distance from the proto-sun since it is too hot and carbon would remain as an atomic gas. We are fortunate to have carbon on Earth. It must have been delivered to the surface via collisions with asteroids and comets that formed in the outer solar system.

Our Moon

As the collisions continue, we think that the dust sticks together because of the electrostatic intermolecular forces between them. They form larger and larger aggregates, eventually aggregating into rocky structures that resemble volcanic pumice stone. As these continue to grow through collisions from millimetres to centimetres to kilometre-sized objects, the solar system becomes filled with literally billions of rocky asteroids. These eventually become massive enough that they begin to attract each other gravitationally and undergo rarer, violent collisions at several kilometres per second. Sometimes the collisions are so powerful that the giant asteroids are shattered into pieces. Within a period of about 100 million years from the formation of the central star, a small number of planetary-sized objects have formed, their masses built up from an enormous number of collisions of smaller rocky bodies. At this late stage the proto-planets themselves sometimes collide together. The collisions are sufficiently violent that they generate temperatures high enough to melt rock, and the early proto-planets are molten hot. Once a proto-planet reaches a size of about one thousand kilometres then gravity will pull all the matter into the simplest equilibrium structure, a sphere.

Some of these late giant impacts have just the right collision geometry and speeds that they can lead to the formation of satellites such as our Moon. The debris from those collisions forms a new disk of material that orbits the proto-planet, and this material subsequently undergoes the same condensation and collisional growth to form a satellite. We know this because each planet has a different composition depending on where in the solar system it was formed; however, our Moon has exactly the same isotopic abundance of elements as the Earth. Indeed, the Moon is made of material that was once inside the Earth!

The material that formed our Moon was literally smashed off the surface of the Earth after a violent collision took place with a proto-planet the size of Mars about 50 million years after the solar system began to form. Astronomers have named the planet with which we collided Theia, after the Greek goddess daughter of the Earth (Gaia) and mother of the Moon goddess Selene. The impact sent the Earth spinning rapidly on its axis, about ten times as fast as today, resulting in a day that was just a few hours long. Furthermore, when our Moon condensed from the material that came from Theia and the Earth, it was ten times closer to the Earth.

There was no life on Earth at this time to witness the spectacular and imposing view of our much closer Moon. In fact, being so much closer, the gravitational force from the Moon would have been a hundred times stronger. This force elongated the Earth slightly and squished it into an egglike shape that always points towards the Moon. As the Earth rotates, the ocean waters "slosh" over these bulges twice each day creating the ocean tides. The average height of the ocean tides today is about one metre, although the complex coastal geometry at the Bay of Fundy in Canada causes tides that are 17 metres high. Four billion years ago, the Moon excited ocean tides that would have been several hundred metres high—tidal waves that would have devastated the oceans' shores. Since the Earth was spinning faster, these would have occurred every few hours—not a good time for life to crawl out of the ocean and explore the land! Planets like the fictitious Pandora (from the movie *Avatar*), which orbit too close to another massive planet would not have such a peaceful environment. Rather, the intense gravitational forces would induce constant earthquakes and volcanoes on its surface.

Our Moon has played a fascinating role in the evolution of life on Earth, both in stabilising our climate and in setting the length of our day. A rotating planet is not perfectly spherical. It is slightly wider at the equator (by about 20 kilometres) since that is where it is rotating fastest, and it bulges as matter is flung slightly outwards by the centripetal force. Many of the planets in our solar system have axes of rotation which are unstable. As a result, they can spin rather chaotically in space as small gravitational perturbations from the larger planets literally cause them to rapidly flip over. Our Earth would do the same if it were not for the presence of our Moon, which provides a stabilising gravitational force on the Earth.

Thanks to our Moon, the rotational axis of the Earth has stayed the same for over four billion years, also resulting in a stable climate over the same timescale. Without our Moon, life on Earth would have developed quite differently, and perhaps complex land-based life forms like elephants and humans would not have evolved at all. A chaotically spinning Earth would suffer dramatic temperature changes on a timescale that can be less than a million years, whereas it took several billion years of evolution on Earth before life developed the ability to be mobile. The first simple life forms evolving in the warm ocean waters and tidal pools near the equator may have suddenly found themselves embedded in thick ice as the Earth tilted over. Life would likely have developed, but perhaps next to the stable environments of the volcanic vents deep under the ocean.

The Earth also squishes the Moon into a slightly flattened egg shape. It is the gravitational pull of these asymmetrical bodies on each other that has caused the Moon to drift away to its current position and the spin of the Earth and Moon to slow down. We know the Moon is moving away from the Earth at a rate of about 4 centimetres per century since the Apollo astronauts left reflecting mirrors on the Moon's surface. By shining a powerful laser pulse at the Moon and measuring the time it takes to receive the reflected signal, we can accurately determine how fast the Moon is receding from the Earth. As a further consequence, the Earth's day has lengthened over time to its current 24 hours. Evidence that the length of the day has become longer over time can be found by observing the growth rings of coral. Some types of coral show growth bands that are sensitive to both the year and the day. By looking at fossilised coral from 400 million years ago, it is possible to count the number of daily growth bands in a year. At that time, the year contained 400 days, indicating a day that was just 22 hours long.

The Moon's rotation has also slowed down, in fact, so much that it is now "tidally locked" to the Earth and rotates exactly once per orbit about the Earth. That is why we only see one side of the Moon. In a few billion years from now the Moon will drift twice as far away, the Earth's day will be about one month long and the Earth will be tidally locked to the Moon—we will only see the Moon from one side of the Earth. Surfers should enjoy the tides now since in the distant future they will only be a few centimetres high and occur every two weeks!

The early bombardment by asteroids was intense, keeping the Earth's surface molten. After about 100 million years, however, the rate of impacts decreased such that the Earth's surface could begin to cool by radiating heat into space. It then formed a solid crust of rock. The first significant rock deposits on Earth are four billion years old; they have traces of sedimentary grains that are smooth and round, caused by their passage down streams and rivers. This implies that water already existed on Earth four billion years ago, as oceans and rivers and an atmosphere with clouds and rain! Some of the water on the Earth's surface was present in the rocky mineral-rich asteroids from which it was formed. Like carbon compounds, some came from comets and asteroids that formed in the outer solar system. The Earth's atmosphere originated from gases trapped within the molten interior, but its composition has changed over time due to the activities of biological life, starting with bacteria, then plants and ultimately humans today.

The origin of planetary systems is a fascinating research area. We have recently carried out an extensive set of supercomputer simulations that model the formation of planets around stars. We found that habitable planets should be very common, with most stars hosting at least one rocky Earth-like planet. Since we could follow the collisional formation history, we could also determine that about 1 in 12 of those planets suffer an early giant impact that can lead to a climate-stabilising satellite Moon. From these simulations we expect that, within our galaxy, there are at least a billion stars which host planets that have stable conditions that could lead to the development of life. The possibility that our galaxy is already filled with life is a wonderful thought that I will discuss further in a coming chapter.

As material collides with the planets, the rate of impacts slows down. However, even today there is still leftover debris orbiting throughout our solar system. Although impacts from larger asteroids are rare, our Earth is struck by a devastating several kilometre-sized object from outer space about every 50 million years. Before we get to the consequences of that, let us review everything we have learned about the history of life and the universe so far.

Chapter 5
One Day

B. Moore, *Elephants in Space*, Astronomers' Universe,
DOI 10.1007/978-3-319-05672-2_5, © Springer International Publishing Switzerland 2014

Following my PhD I was fortunate to be awarded a fellowship that I could choose to take almost anywhere in the world to continue my research in cosmology. I was destined for the University of California, Berkeley, which I chose for its famous scientists and its proximity to Yosemite Valley whose spectacular granite walls I longed to climb. My then wife, Gabrielle, and I set off driving to Yosemite after lunch on the first weekend after arriving in California. It was only an inch on the map, but I was not used to the vastness of things in America and did not realise that this "inch" would take five hours to drive. The sheer granite walls stretch as high as a vertical kilometre above the valley floor, which had been home to the Ahwahnee Indians for several thousand years until they were displaced by the California gold rush in 1848. Gold is a strange thing. It is one of the few elements to be found in almost pure form –gold nuggets. When stars explode, all the elements are ejected into space as individual atoms. So how come gold is the one element that is found on Earth in large lumps? I was wondering about the answer to this question five years later while suspended from a rope halfway up Half Dome in Yosemite in an attempt to free-climb its spectacular North-West face. As a student or post-doctoral fellow, you have the freedom to think with none of the time-consuming bureaucratic distractions that come with professorships. Much of my time was spent with eminent colleagues in Café Strada at the corner of the Berkeley campus. Beneath the sun-shielding palm trees, we discussed our research and contemplated deep questions while drinking cappuccinos you could buy for a dollar. We were entertained by a constant stream of interesting Berkeley folk passing by, the leftovers of a previous generation of colourful, peace-loving hippies who survived the "electric Kool-Aid acid tests" of the 1960s. Any free time was spent climbing, a passion ignited and shared by Gabrielle. Our adventures could fill a book. Indeed, I could tell tales of our encounters with bears and wolverines in Yosemite, being chased for miles by wolves in the Dolomites or crossing the Indus River in an orange crate on our journey deep into the Karakoram Range. . .Life is good.

It is difficult to make sense of the development of our universe owing to the enormous range of timescales involved. To make this more comprehensible, let us rescale our unit of time by compressing the entire 13.8-billion-year history of the universe into exactly one solar day—that is into a 24-hour timeline. This means that every hour that passes on our new clock represents 575 million years of history, each minute corresponds to almost 10 million years and each second is 160,000 years. I will describe some of the most important moments in the formation of cosmic structures, alongside the development of life on our planet. We can then at least appreciate the relative times between events. This will establish the place of our species in the grand scheme of things.

00:00:00

Bang on midnight: Space, matter and energy appear, time begins and the universe bursts into existence using a net energy budget of zero. We are so used to cause and effect—to the idea that "nothing happens unless something makes it happen"—that we naturally want to ask How? Why? and What? This logical principle governs our everyday experience and drives our search for a mechanism and a reason as to why the universe began. Did nothing exist before the big bang? Surely something cannot come from nothing? But it is a remarkable fact that the total energy of the universe is actually zero. As my colleague Lawrence Krauss emphasises in his book "A universe from nothing", if you add up the positive energy in all the particles, it is exactly equal to the negative energy of the gravitational forces and vacuum energy. So the universe does in fact consist of essentially "nothing". Fortunately for us, that nothingness is split into equal positive and negative parts of matter and energy. We are one of the consequences of all of this. But what is our destiny?

00:00:02.37

Zero hours, zero minutes, 2.37 seconds after the start of our brand-new day, the cosmic microwave background radiation emerges and fills the universe. An awful lot of interesting particle physics has already taken place, including nucleosynthesis and the formation of atomic nuclei. There are no stars in the universe, no planets, no life. A period of darkness ensues, although gravity continues its relentless task of gathering together the massive particles: The structures in the universe are slowly forming.

00:05:00

By 5 minutes after midnight the first complex structures have formed—the first dark matter haloes; the building blocks of galaxies are assembling, driven by the force of gravity. They emerge from the fluctuations that are present in the "primordial soup", a phrase used to describe the conditions during the earliest era of our universe. These structures grow by merging together in violent collisions, forming ever larger and larger dark structures.

00:20:00

It is just 20 minutes after midnight and the very first star appears, a pinpoint of light illuminating a tiny region of the universe. Then another, and another and soon billions of early stars begin shining throughout the universe. These first stars are a little different from stars like the Sun since they are made from a volatile mixture of pure hydrogen and helium with a sprinkling of lithium. At this time there are no elements in the periodic table heavier than lithium; therefore, rocky planets like the Earth cannot yet exist. The rate of nuclear fusion at the centre of each of this first generation of stars is much higher than occurs in the Sun and leads to the rapid formation of many new elements. They live for a brief moment, emerging shining and then dying within about six seconds on our one day timescale (a million years in real time). They die a spectacular death, exploding as supernovae—cosmic fireworks on the grandest scale. The newly created elements of carbon, oxygen, nitrogen. . .and elements from across the whole periodic table are scattered throughout the universe, eventually reassembling within new stars and forming the first planets.

01:00:00

The first galaxies appear one hour after midnight, collections of millions or more stars that form at the centres of the now massive dark matter structures. These are the most distant objects that we can see with our telescopes, and they lie close to the edge of our visible universe. Their starlight has taken a long, long time to reach us, a journey that occupies the rest of the day. The images we see are snapshots of how those galaxies appeared shortly after they formed, not long after the beginning of the universe.

03:30:00

The universe already contains trillions of stars which fill a vast number of galaxies. They form and evolve like our own Sun, shining for the rest of the day and beyond. Based on the timescale needed for life to emerge on Earth, by 3:30 am there has already been plenty of time for life to have developed throughout our universe long before us. If we were actually alone in our galaxy, it would indeed be very strange.

09:00:00

It is breakfast time, and it has taken nine hours to make half of the stars in our galaxy. Already 50 billion stars have formed and a similar number will appear over the rest of the day. The stars and gas have arranged themselves into a huge rotating disclike structure—our Milky Way. At this time the Sun has still not appeared, yet generations of stars are being born, the most massive of which explode as supernovae, lighting up the galaxy at a rate of tens of thousands per second.

16:00:00

It has been a busy day and is time for afternoon tea at 4 pm. Another star forms, no different from all the others, but it is special because it is our Sun and it provides the energy to sustain life on Earth. The Sun and its planets lie 24,000 light years from the centre of our galaxy in one of its spiral arms. At this time it is beginning its first lap of a total of 20 that it will make around the Milky Way by the end of the day, just like the planets orbit the Sun but on a much longer timescale.

16:05:00

Minutes after our Sun ignites and begins nuclear fusion, our planetary system is already in place. The cloud of gas and dust which collapses to form the Sun ends up in a dense rotating disk consisting mainly of hydrogen and helium gas. But it also contains all the elements in the periodic table that were made at the centres of stars. This atomic material forms the planets, including Earth, by gravitational collisions. These impacts of early proto-planets impart so much energy that Earth becomes a large sphere of molten rock, consisting of about 35 percent iron, 30 percent oxygen, 15 percent silicon and 13 percent magnesium, with the remaining 7 percent coming from all the other elements. At this time there is no water on the planet's surface, and it has no atmosphere.

16:10:00

A late giant impact occurs between our proto-Earth and Theia—the name given to the colliding body. The debris it scatters in orbit around the planet coalesces to form our Moon.

16:20:00

The surface of the Earth has cooled sufficiently to allow a thin crust capable of holding liquid water without it boiling away. At this time the solar system still contains numerous comets and icy asteroids which are colliding with the Earth—continuously supplying water to the planet's surface, albeit through shattering collisions. Water is essential for the rapid buildup of life on Earth in the form that we recognise and understand. It provides a stable and dense medium within which complex molecules can form. This period is called the late heavy bombardment. Evidence for it comes from the dating and composition of Moon rocks and craters. Unlike the surface of our planet, which is constantly changing through climate effects of erosion and geological effects of plate tectonics, the Moon has no atmosphere. Its crust is ancient and its surface contains a record of the earliest phases of the formation of our solar system.

17:00:00

It is time for an evening banquet to celebrate the achievements so far. The essential ingredients and environment for life to develop are in place: an energy source, a host of atomic elements that can bind together in ever more intricate forms and a liquid environment that is stable for long periods within which complex replicating molecules can form. Life begins to emerge like a phoenix from the ashes of long-dead stars.

It is difficult to reconstruct the earliest phase in the development of life. Fossilised remnants of initial life forms are lacking since they had no hard shells or bones. However, evidence shows that single-cell prokaryotic microbes appear about now, 500 million actual years after the formation of the Earth. These are the earliest forms of biological life—microbes that can replicate and metabolise and which contain the essential ingredients for life: ribonucleic acid (RNA), deoxyribonucleic acid (DNA) and proteins (consisting of amino acids).

19:30:00

Cyanobacteria begin to oxygenate the Earth's atmosphere through the type of photosynthesis we recognise today that releases oxygen as a by-product. This leads to the largest extinction event in the history of our planet as most existing life forms cannot survive in the new oxygen-rich conditions. Cyanobacteria trigger one of the most violent climate fluctuations on record through an early greenhouse-type catastrophe in which the Earth is almost entirely frozen and glaciated for the longest period in our history.

20:00:00

Eukaryotes appear, the first complex cells, with a nucleus enclosed in a membrane. The emergence of these single-cell structures is a big step in evolution—they form the basic structure of all the animal and plant life on our planet today.

20:30:00

Multicellular life has evolved independently over the history of the Earth in both plants and animals. The first large organisms that appear at half past eight in the evening of our 24-hour day are simple soft structures with no bones or hard body parts.

23:05:00

It takes a further 2.5 hours of evolution before animals appear, about an hour before the end of this day. Animals are eukaryotic multicellular forms, which distinguishes them from unicellular bacteria. They digest food internally, separating them from plants and algae, and they can move independently! Evidence for centimetre-long wormlike creatures appears at this time, although the relatives of most known animal species emerge during the next half hour. This period is known as the Cambrian explosion; during this time creatures evolve that manage their movements and habits with a central control system—a brain! It is time for a cocktail and to sit back and watch the rapid development. Life is indeed good.

23:07:00

The first vertebrates appear: fish with backbones.

23:12:00

The first land plants begin to cover the Earth's surface, which has been rocky and barren for the previous seven hours.

23:20:00

Sharks appear and begin to terrorise the oceans. They continue to do so until midnight, becoming one of the oldest surviving species.

23:24:00

Amphibian creatures begin to explore the dry land and use land plants for food.

23:30:00

There are only 30 minutes left until the end of the day, and the first air-breathing reptilian land animals begin to walk the Earth.

23:35:00

Dinosaurs take over as the dominant species at the very top of the food chain.

23:40:00

The first milk-producing animals appear, leading to over five thousand mammalian species by the end of the day.

23:46:00

A branch of dinosaurs develop feathered wings and take over the skies: the ancestors of birds.

23:48:00

Flowering plants appear, bringing colour to our planet. The colours attract birds and insects, which aids the plants' pollination and transport of seeds across the Earth.

23:55:00

A giant asteroid hits the Earth at 5 minutes to midnight. The fossil record and chemical composition of strata laid down over time indicate that giant impacts from asteroids were common and happened about every 5 minutes on this timescale. Over 90 percent of all species present are wiped out, including the dinosaurs. They have had a lengthy rule at the top of the food chain, but they have not managed to evolve the capability to progress further, to develop tools or to use fire. They have lived as other animals have lived, a day-to-day existence of eating, procreating and sleeping.

23:59:40

The surviving life adapted and evolved, and biodiversity increased again. With only 20 seconds left to go until the day ends, Australopithecines diverge from other primates and begin to explore the world on two feet.

23:59:58.85

At one second to midnight, *Homo sapiens* emerge as a distinct species and begin their journey out of Africa. There is no turning back.

23:59:59.80

In the final two tenths of a second after the last great ice age, agriculture becomes widespread. Organised society appears and humans now have time for activities other than the daily search for food. Now the music and fun begin—let the party start!

23:59:59.95

Organised societies leave their spectacular mark with the construction of long-lasting monuments epitomised by Stonehenge in England and the Pyramids in Egypt. Both require advanced construction techniques and hundreds of workers over many years. On our 24-hour clock these milestones in human achievement happen just 50 milliseconds before midnight.

23:59:59.99

The ancient Greeks construct the first mechanical computer, the Antikythera mechanism.

23:59:59.999

At 1 millisecond to midnight Albert Einstein is born.

24:00:00

A long and eventful day has come to an end, and all recorded human history has taken place in the final second. The Earth's population has just reached 7.0 billion people. Our rule at the top of the food chain is several hundred times shorter than that held by the dinosaurs, yet in that relatively short time we have developed the technology to protect ourselves from the same death by asteroids that led to the extinction of the dinosaurs. We have also developed the technology to wipe out most, if not all, living things on our planet. So it is a good time to consider whether or not we will still be here one second after midnight. Contemplate for a moment whether or not *Homo sapiens* could outlive all previous species by surviving for a further hour on this timescale.

One solar day on this timescale lasts just 17 nanoseconds. In this time 400 million people have sex, 400,000 babies are born and 200,000 people die, 30,000 of them from starvation. Over a trillion queries are made on Google and even more emails are sent. Five hundred square kilometres of forest are destroyed and 100 unique species become extinct. If our own species can avoid self-extinction and survive for a mere one minute after midnight at the beginning of tomorrow, then in this short relative time we could fill our entire galaxy with life.

Before looking at how we can achieve this I would like to first look at how it is that our species could acquire all of this remarkable knowledge. In the next chapter I would like to focus on the brain, our tool for understanding that seems to surpass that of any other species on the planet.

Chapter 6
The Brain

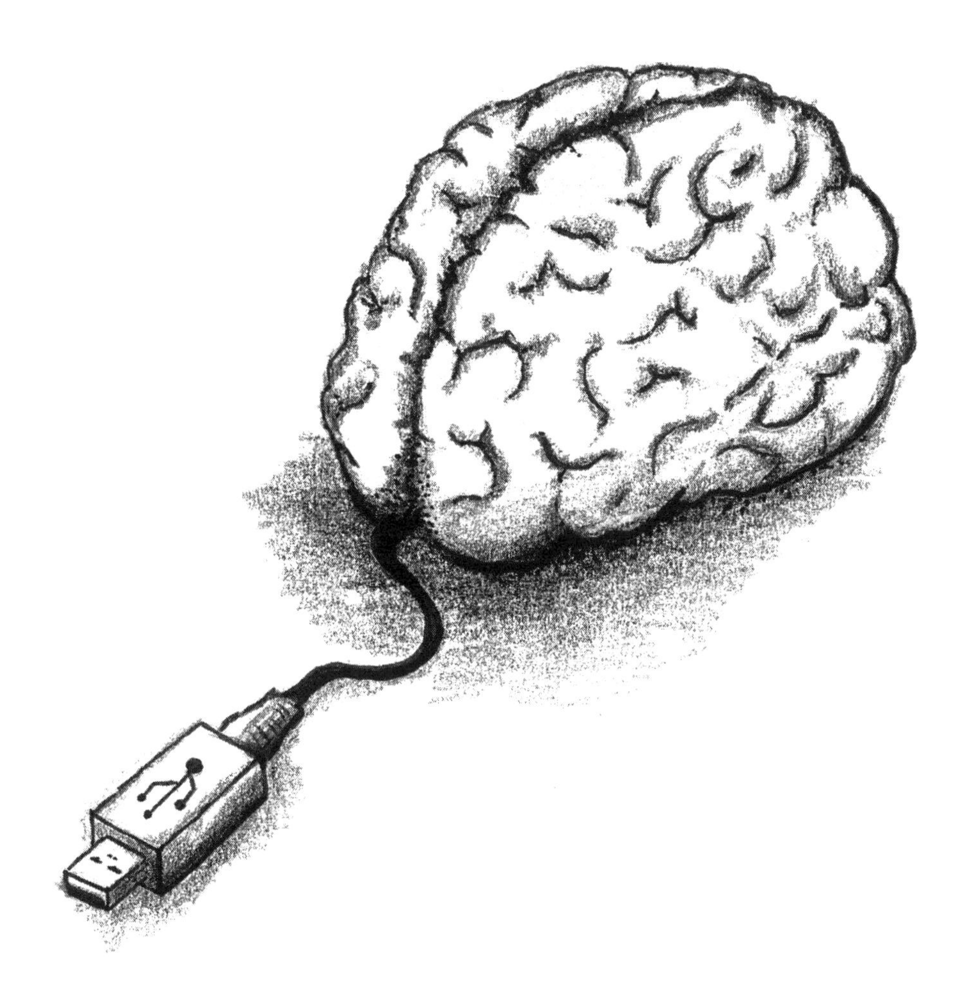

B. Moore, *Elephants in Space*, Astronomers' Universe,
DOI 10.1007/978-3-319-05672-2_6, © Springer International Publishing Switzerland 2014

As a young theoretical astrophysicist I became more and more interested in using powerful computers to simulate our universe. But I was also rather envious of my observational colleagues who kept disappearing to spend days and nights at spectacular mountain-top observatories in exotic places like Paranal in Chile. I accompanied a colleague to a place that shall remain nameless (for reasons that will soon become apparent) to observe distant galaxies through a powerful telescope. It was not what I expected. A modern observatory is a high-tech laboratory. The operator sits in a control room away from the actual instrument so that the temperature changes in the telescope dome are minimised. The user does not even look into the lens of a telescope anymore. The coordinates are entered and the telescope moves into the correct place and begins to track the object as it moves across the night sky. The telescope collects as many photons as possible and focuses them onto a detector like that in a digital camera, but much larger and more sensitive, and it is immersed in a bath of liquid nitrogen to reduce thermal noise. The detector counts incoming photons literally one at a time and records their position and energy. The images of faint distant galaxies, invisible to the naked eye, slowly appear on the computer screen. After an exhausting week of sleepless nights during which time we could not even open the dome because of the incessant rain, the clouds began to clear. The conditions were still too poor to allow readings of useful data, so I thought it would be fun to point a really big telescope at the Moon. Well, it turns out that the Moon is very bright. Much, much brighter than the faraway things at which the telescope was designed to look: a thousand trillion times brighter, in fact. The number of photons landing on the detector caused it to instantaneously boil the liquid nitrogen. For a brief instant, I saw in exquisite detail the shear mountainous walls that rose five kilometres high above a deep impact crater on the Moon. Then the screen blanked out as the system overheated and died. That was the end of observing for the night. We slipped away quietly to the sleeping quarters, hoping no one would notice. The next morning we returned to survey the damage and heard a peculiar grinding noise coming from inside the dome. We had forgotten to instruct the telescope to stop observing, and it had continued to track a fixed point on the night sky which had long since passed beneath the horizon. The telescope was upside down pointing into the ground, the drive gears whirring madly, the mirror barely attached to its frame. I decided to stick to theoretical cosmology and have not been observing since.

Over the past 20 years I have thought long and hard about various aspects of our physical universe. I think I have a reasonable understanding of how structure emerged from the big bang, how our galaxy and solar system formed, how the Sun shines and when it will die, how neutron stars and black holes form. However, the most amazing and dauntingly complex thing that I have come across in our entire universe is the human brain. The brain is a product of countless generations of incremental changes, a road of optimisation that functions in the face of the conditions provided by the Sun and the environment on Earth. It is an elaborate molecular computer capable of processing and storing information at a rate that exceeds that of the world's largest supercomputers. It interfaces with the physical surroundings via at least ten different senses, and the amount of information that it simultaneously processes is simply staggering. It stores our memories, makes our decisions and holds our consciousness.

Our level of understanding of how consciousness works is similar to our understanding of the first millionth of a second of the evolution of our universe—not good! However, science has made huge progress over recent years, and we are beginning to formulate an understanding of how at least some parts of our brain function. The ancient Greeks dissected human nerves and suggested that visual rays caused sight. Hippocrates maintained that the brain is part of sensation and the seat of intelligence. The Greeks questioned how our thoughts work, a question we still ask today. I would love to know how, when I close my eyes, I can bring a vast number of stored images and memories into my consciousness, and manipulate them at will to create imaginary scenes and sensory experiences. In the past hundred years there has been an explosion in our understanding of the brain equal to that in our knowledge of our place in space and in time in this universe. This increase in knowledge has been driven by our understanding of how matter behaves at the atomic and molecular level. And with our brain, we have the possibility to perform neuro-physiological and psychological experiments as well as direct mapping and monitoring of its neuronal network.

The power of the brain is remarkable. The coordination and data processing required to perform a simple task, such as taking a drink, is simply staggering: beginning with some stimulus or thought that triggers the idea of having a glass of champagne, interpreting the stream of visual information while coordinating hand movements, monitoring touch and temperature, controlling balance, and finally synchronising swallowing and breathing. All of this while at the same time maintaining a conversation and wondering why those bubbles keep rising from the bottom of the glass. Finally, once the drink is down, we no longer have to think about how to process it. That is done automatically through our autonomous nervous system and responses to chemical reactions that take place inside our bodies, from the release of the appropriate digestive juices to the breakdown of the liquid into its molecular components, which are then sorted and distributed throughout our bodies.

For some reason we *Homo sapiens* have gone beyond other species in our mental capabilities, and our capacity for technological and scientific advancement has increased exponentially as a consequence. I want to describe to you a little of what we know about how our brains and senses work, what our brains might be capable of and how they might evolve in the future. I would like to compare our brains with our digital supercomputers and to set the scene for later discussions on the longevity of life and the question, Can biological life, or artificial life, survive for eternity within our universe? The brain, like a computer, requires energy to function. Each decision we make, each memory we store, requires something to change its state as a result of an exterior action or event. However we store a memory, be it with pen and paper, inside a transistor or ultimately within a molecular structure, we need to change the state of something that already exists. Changing the state of a physical object requires an external force—some energy through cause and effect. As we will see in the later chapters, energy is a rapidly decreasing resource. If we wish to continue to think, we must develop the intelligence and technology to extract and use that energy.

The Human Brain

An adult human brain contains about 100 billion neurons which receive and send information via electrochemical signals. An individual neuron has a complex internal structure that contains genetic information, can synthesise proteins and produce energy to fuel cellular activities. A typical size is about 0.01 millimetres—you could only fit around ten of them on this full stop. Neurons receive information via a complex treelike structure of thin protoplasmic wires which connect up to 10,000 neurons together. Our brain has about 100 trillion such connections. Whereas each neuron can have thousands of inputs, it only has one output along its axon fibre. After leaving the neuron, this fibre branches and transmits a signal to numerous connected neurons or cells in our bodies. The longest are the motoneurons, which extend for a metre or more from the spinal cord to your toes. You could fit hundreds of axons within a single strand of a fine human hair, which is itself only about 0.01 millimetres in diameter.

The wiring of our neural network fills most of the volume in the brain—if you could stretch it all out in a line it would extend to a length of over 100,000 kilometres; two-and-a-half times around the Earth! The network in women is 20,000 kilometres longer than that of men; although the reason is unknown, one could certainly argue that women are complex beings. When our nerve endings detect a disturbance, such as increased pressure or temperature, they generate a tiny electrical current that is passed along the axon via chemical voltage changes between a long line of neighbouring charged ions. Information flows down the axon without loss of signal strength at speeds from 1 to 100 metres per second, depending on its thickness. It reaches the neurons in the brain which act like tiny volt-metres by maintaining a potential difference across their membrane. If a neuron receives a certain number of incoming signals within a given time, it will fire a response via its axon through numerous synaptic junctions that pass the signal to other neurons. A given neuron can either pass on the information to many more neurons or ignore it, depending on the cumulative effect of the incoming signals. This whole process sets our reaction times to external events. It is not like an electrical wire which communicates information at the speed of light; rather, each molecule is influenced by the next molecule like a line of dominoes. As a result, it takes about 200 milliseconds for an average person to respond to an audio or visual stimulus.

Neurons maintain a voltage gradient across their cell membranes through different concentrations of ions, electrically charged particles of sodium, potassium and calcium, rather like a battery. If the voltage across the membrane changes by a large enough amount in a certain time interval, it may transmit a signal along its axon. Individual neurons (or groups of neurons) fire bursts of electrochemical millivolt pulses lasting less than a thousandth of a second which travel along their axons to their connected neurons at the synaptic junctions. When the same connections receive continued stimulation via responses from our senses, a particular pathway between neurons becomes easier to traverse since the synapses undergo a chemical change on repeated use. An imprint, or memory, is left embossed on the neural network of the brain. As we accumulate knowledge and grow older, we rewire our neural network, forming new connections and ever more complex networks of wiring.

Our Senses

The neurons with their axons are also part of our nervous system which interfaces with our senses. They detect disturbances from the outside world and convert them into electrochemical signals to carry to the brain for interpretation. The nerves are cylindrical bunches of fibres that extend from the brain and spinal cord, then branch repeatedly and stretch to all parts of the body. Nerves are large enough to have been recognised by the ancient Greeks, although their structure and purpose could not be determined until the advent of the microscope. Although we understand the individual components of our senses quite well, we know little about how the entire neural network functions to make decisions and to control our actions and thoughts. Neither do we know the full details of how our brain stores and reads memories. A working brain cannot simply be dissected to see how it is working; there are far too many pieces and its components are at the molecular scale. Understanding how memory works is one of the key areas in neuroscience research, and psychological tests, chemical tagging and brainwave mapping are among the tools that can be used to study how the brain functions.

Our memories are filled with input from our senses which allows us to perceive the world around us. These perceptions are formed inside the brain, and their reality has kept generations of philosophers arguing. Our sensory system—sight, smell, touch, hearing and so on—operates at the molecular level. Our senses detect and measure changes in our exterior environment which trigger chemical reactions around our nerve endings. This causes an electrical signal to travel along the nerves carrying information of the event to the brain which then processes the information. It compares it to a repertoire of previous events to decide how to respond. The brain learns most of these abilities; it is not something with which we are born (anyone watching a baby develop its reaction to events can see this). They learn quickly that hot things are bad, that sharp things cause pain, to recognise their parents and that breasts produce food.

Smell, touch, hearing and taste all work in a similar way, through the stimulus of special receptor neurons at the nerve endings. Together with sight, these are our main senses, but we can also sense temperature, motion, pain, balance and acceleration. Some of the senses, such as balance or hearing, use a mechanical device to interface with our brains. Some animals are also sensitive to magnetism—birds can sense the horizontal and vertical components of the Earth's magnetic field, which helps them navigate during their long migrations around the world. Scanning images on Google Earth revealed that cows and deer preferentially orient their bodies according to the Earth's magnetic field when grazing or resting. When in the vicinity of high-voltage power lines, the animals were randomly oriented, presumably because the magnetic field strength generated by electricity cables overwhelmed their magnetosensory system. Living creatures have evolved with the capabilities to sense the physical conditions around them because it enables their survival in a competitive world.

Vision is perhaps our most favoured and important sense: Our eyes are another remarkable product of hundreds of millions of years of evolution. Over 95 percent of animals have eyes, and there is great diversity in how they work. The human eye is an optical lens that focuses light onto an enormous number of receptors at the back of the eyeball. The retina, where the image of the world is recorded, is covered by 120 million very sensitive photoreceptor rods and about 6 million colour-sensitive cones. This is part of our optic nerve, which has over one million fibres that transport the information to the visual cortex. This part of the brain processes the visual image and is the largest network of connected neurons inside the brain. When about 100 photons of light have hit one of the sensors on the optic nerve, their energy is converted into an electrical impulse that travels to the brain, which "interprets" the information as a colour of a certain luminosity (brightness) at a specific location. Colour is the means by which the energy information of the incoming photons is recognised.

The image that is focussed onto the back of the eye is actually upside down—it is just like looking through a magnifying glass at arm's length. The brain automatically inverts the image so we see the world the correct way round. To prove this and to demonstrate the remarkable learning abilities of our brain, as an experiment, people were given glasses that focussed the incoming images the other way up. This world appeared upside down at first, but after a few days the subjects' brains had automatically turned the image the right way up.

Our eyes are sensitive to light with wavelengths from about 370 to 740 nanometres, that is from the violet part of the spectrum to red. We are able to see objects of different colours since the Sun's white light contains a mixture of all colours and different materials reflect or absorb different parts of the colour spectrum. An apple looks "red" because it absorbs most of the other colours and reflects those photons in the red part of the spectrum. Tests with colour charts reveal that the eye/brain combination has the ability to distinguish ten million different colours! At any instant, a stationary eye can detect a contrast ratio of 100 to 1, but as the eyes move and adjust their exposure, within about 30 minutes they are able to detect a contrast ratio of one million to one. Contrast is a measure of the brightest to the darkest shades that are detectable. In principle the eye can detect a single photon of light, although various molecular filters trigger a signal to the brain once a receptor has received about ten photons within about 100 milliseconds. Our eyes function like a digital video camera that takes a continuous stream of information which is then processed, analysed and stored or discarded in real time by our brains. In terms of "pixels" (the smallest visible unit of which an image is composed), the eye has about the equivalent resolution of a good quality 10-megapixel digital camera. Every few seconds, they deliver to the brain as much information as can be stored on a compact disc.

Evolution has followed a haphazard path, resulting in a remarkable diversity in living things. Each species has evolved components according to needs determined by the environment in which it lives. Our smell is millions of times less sensitive than that of a bear, which can smell food 10 miles away. Bats can hear three octaves higher than we can. Our eyes pale into insignificance compared with the colour range and sensitivity of those of the incredible mantis shrimp, which can even see the world in polarised light. We can only run a third of the speed of a cheetah and swim a tenth as fast as a sailfish. I wonder why an "optimal creature" has not evolved to have all of these advanced sensory and physical attributes.

Memory

When the brain receives a certain stimulus, for example when you look at someone for the first time, a vast network of neural activity takes place in a particular sequence. This sequence becomes a short-term memory that lasts less than a minute for most people. If you never see this person again, neither the particular sequence of activity nor the pattern of activity across your neural network will ever recur and the "memory" is forgotten. But if you see and think about that person regularly, the pattern recurs and the sequence is imprinted in your neural network. This process is facilitated by the part of the brain called the hippocampus, which consolidates long-term memories. Memories are formed and held through specific connectivity and communication patterns between neurons that are strengthened by repetition. I will discuss later how information can be stored in our neural network by comparing it to a digital computer that uses switches and logic gates.

Research findings show that our memories work in this way. Recent experiments on the specific area of the brain that processes learning established that specific memories can be detected and permanently erased. By teaching a certain response to a musical note, a mouse was trained to fear an event happening. After the training, even if the event did not happen when the note was played, the mouse would freeze and wait for the expected event. The neurons in its brain that were activated during the experiment were highlighted and marked using chemicals that respond to neural activity. The neurons that became active once the note was played were subsequently destroyed, after which the mouse completely forgot to be scared when it heard the tone. It could function normally afterwards and could even relearn the same fear response albeit programmed onto a different set of neurons. Chemically tagging active neurons makes it possible to construct a three-dimensional map of the neural activity in a brain. Ultimately, this technique could allow us to decode and reverse-engineer how our brains actually think.

Confirmation that how we react to and interpret sensory signals is a learned phenomenon and that our senses are "unconnected" was published in 2010 by Pawan Sinha. In the seventeenth century, the Irish scientist and philosopher William Molyneux posed the following question: Could a person who was blind from birth, who held a cube and a sphere and who could tell the difference between the two, identify each through vision alone upon obtaining sight for the first time? This remained one of the foremost questions in the philosophy of the mind for three hundred years because there was no good way to test it until recently, when scientists restored sight to children in India through a relatively simple medical intervention. Before the medical procedure, the blind children were given objects to hold and become familiar with. On being able to see for the first time, the children could not distinguish visually between objects that they could previously identify by touch. Within a matter of days, however, they could distinguish the objects by sight alone.

The fact that our senses need to be trained and that we can disable and erase a specific memory is difficult to reconcile with the concept of a "spirit" or 'life force' that holds a soul and our memories. Everything we understand about the way the brain functions is consistent with the idea that it is a sophisticated and complex molecular computer. Perhaps in the womb, our small embryos develop a brain filled with neurons that are connected together, somewhat haphazardly. Then, as we explore our world and learn responses to certain stimuli, we imprint and tune the circuitry that becomes our memory and our learned responses. We are born with a sparsely filled canvas that we paint with knowledge and memories.

How a Computer Works

I will discuss brain size and intelligence shortly, but first I would like to compare the capabilities of our brain with modern computers. While the evidence shows that our brains are an outcome of natural selection, computers are the result of intelligent design. I do not claim that computers are equal to the human brain, or that they even function in the same way. But it is an achievement of our brain that we have designed something that may actually be compared to it, albeit rather crudely. The interesting and scary fact about the machines that we can design and construct is that their sophistication and power are increasing exponentially. The consequence is that during a timescale over which our species will evolve barely noticeably, machines will be constructed that will dwarf the computational abilities of the human brain.

In 1854 the English mathematician George Boole showed how algebra could be adapted to show logic instead of arithmetic. The variables in the equations (the numbers) can only take on two values, true or false. The mathematical operators, like plus and minus, are replaced with rules of conjunction (and) and disjunction (or). Then, in 1938, the American mathematician Claude Shannon, the inventor of information theory, showed that electronic circuits could be used to perform sequences of operations using switches that could be in either an "on" or "off" state to represent the true and false outcomes of logic. The first computers used valves as the switches whereas today we use tiny transistors. A single transistor has two input wires and one output wire that are connected to a tiny piece of semi-conducting material such as silicon or germanium. A transistor can act either as an amplifier or a switch. In amplifier mode, a small input current results in a much larger output current. In switch mode, one of the inputs is kept at five volts. If any current is passed to the other input, the transistor will output five volts; otherwise, it does nothing. That sounds simple, so how can computers make such complex calculations? The trick is to connect together several transistors to form a more complex switch, known as a logic gate. There are several logic gates, with names such as AND, OR, NOT and NAND. However, it was shown by mathematicians as far back as 1880 that all logic gates can be made from combinations of NAND gates. An NAND gate returns an output signal for any combination of input signals except the case in which two input signals are present, in which case it outputs nothing, i.e. "not and".

Combinations of logic gates can perform mathematical operations at any level of sophistication. This was already proven mathematically by Alan Turing in 1930, long before computers were constructed. He considered the potential computational power of a machine operating with a single logic gate which performed operations on an infinite stream of input data. The neurons in the human brain function in a way similar to logic gates. However, the logic gates in a digital computer work with fixed voltages and either do or do not output a voltage, resulting in two states: "on" or "off". Neurons perform logic operations based on multiple voltage inputs and may or may not yield an output signal. They are analogue devices that operate using variable chemical and resistivity changes—they can process signals of different strengths and signals that enter the neuron at different rates. But just like a transistor, the output signal from the neuron is at constant voltage strength.

Computers store and read information in many different ways, and the technology is always changing. Many devices also use transistors to store data, such as on portable memory sticks. These come in different sizes. The smallest microSD devices that you use in your telephone are about two millimetres thick and smaller than a postage stamp, yet they can currently store up to 64 gigabytes of data. Each "byte" is one character that is stored using eight "bits" of information. A single bit is the basic unit of binary information and is represented as either a 1 or a 0. The eight bits allow a binary number from 0 to 255 to be formed. In binary notation the number 255 is written 11111111. Larger numbers would require using extra bits. With eight bits, each of those 256 numbers can be mapped into a character set containing the letters and symbols of language for manipulating and storing words. So, this tiny little microSD card can hold 64 billion characters, or about a hundred thousand books!

The transistors can be in two states, on or off, which represents 1 or 0, by either storing a charge or not storing a charge. It takes four transistors to store a character in binary format, so there are 256 billion transistors inside that tiny microSD card! Some of the latest transistors to be designed can process up to four levels of input current, allowing each transistor to store more than one piece of information, perhaps even closer to how a neuron works. Information can be read or written to a memory stick up to a million times before the circuitry degrades and the transistors no longer hold their charges. Our memories can last a lifetime, and there is no known limit on the number of times we can access them.

Scientists are working hard to understand how to use molecular computing. Individual logic gates constructed from single molecules have already been successfully made by researchers at IBM. In the near future, it might be possible to store the entire contents of YouTube on a device no larger than your mobile phone.

Everything you copy onto a memory stick is converted into a binary format of 1's and 0's. An image is encoded pixel by pixel, with the coordinates, colour value, intensity and so on all stored as binary numbers. If you record the image as "true colour", which gives the possibility of over 16 million individual colours, then this requires 24 bits of information for each pixel of the image. A 10-megapixel image requires 10 million times 24 bits of information to be stored, which is about 30 megabytes of storage for a raw, uncompressed image. This image can be compressed to a much smaller size, but the final image quality depends on how much information you are willing to sacrifice. The data rate from your eyes to your brain is therefore huge—information from a million pixels flows at multiple frames per second into your brain, that's about ten million bits per second of information—equivalent to a very fast internet connection. Our brains process that stream of data in real time and store the information using sophisticated data compression techniques.

Brain Power

Just how powerful is the human brain, and how does it compare with our digital computers? Information is transferred when something changes its state as a consequence of some external signal. Consequently, by considering how our neurons function, we can make a crude estimate of the rate at which our brains can process information.

A typical neuron sends information at a rate of about 100 times each second, and each neuron is connected to an average of about one thousand other neurons. The neurons seem to function similarly to a logic gate, in which case we can assume that every neuron is capable of at least one calculation for each input/output signal. The human brain has 100 billion neurons. Therefore, it has the potential to make a phenomenal 10^{16} calculations per second if they are all used in parallel! The memory capacity of the brain is a bit trickier to calculate since the details of how we store a bit of information is not well understood. The evidence suggests that we store information through the synaptic connections—the junctions between the wiring of the neurons. If each synapse holds one bit of information, our 100 trillion synapses should be capable of holding about 100 terabytes of information. These are very approximate "order of magnitude" estimates, but they illustrate the incredible power and memory capacities of a single human brain.

We can compare this with the central processing unit, or CPU, of a computer which works by switching on and off its transistors at a rate determined by the "clock speed" (or frequency). The first computers in the 1960s operated at about 100 cycles of on and off per second and thus performed a maximum of 100 calculations per second. However, computers have been getting faster, components have been shrinking in size and clock speed has been steadily increasing. The exponential increase in computer speed with time was first noted by Gordon Moore, the founder of Intel, and the trend of a doubling in speed every 3 years has been maintained for the past 40 years. In 2014 the fastest CPU has a clock speed of several gigahertz, which means several billion cycles each second (one hertz equals a single cycle of on and off each second). For each cycle of the clock, a modern CPU can also perform several individual calculations, resulting in about ten billion calculations per second. So, your brain is about a million times faster than your home computer, both in its calculating power and short-term memory capacity!

Today, individual computer CPUs are not getting much faster; rather, each new generation may have multiple “cores”, or compute units, which operate in parallel. A CPU that has eight individual cores, each performing a set of calculations during each clock cycle, can boost the speed of a computer by a factor of eight. Furthermore, individual CPUs can also be connected together to work in parallel. In 2013, the world’s fastest supercomputer is the Chinese Tianhe-2 (which is Mandarin for “sky river”, meaning our galaxy), which is made up of about 3,120,000 compute cores. These are all connected in parallel using a high-speed data network that allows different CPUs to communicate with each other. It can perform over 10^{16} calculations each second and has 262 terabytes of short-term memory. Its computational power and memory-storage capabilities are comparable to that of the human brain.

Our computers are far less efficient than the brain, though. The Tianhe-2 supercomputer occupies a floor space the size of a football field. It needs 200 people for its maintenance and operation and requires 17 megawatts of power to function! Compare this to our brain, which fits neatly into a 16-centimetre cavity, maintains itself for decades and requires just 20 watts of power to function. Power loss to a supercomputer causes a complete system crash instantly. Our brain can survive for about 10 minutes without power, in the form of oxygen deprivation, but longer than that and our circuitry suffers irreversible damage.

There are many tasks that a computer can do much faster than a human and vice versa. A normal laptop computer can do millions of mathematical calculations and commit them to long-term memory in just one second. I can do about one simple math calculation in the same time and commit it to my short-term memory. The brain is capable of processing much higher amounts of information, but calculating a million mathematical functions in one second has never been important for our evolution, and our brains have not adapted to include this skill. On the other hand, our abilities at tasks such as pattern recognition, for example identifying a person in a crowd, currently exceed the capabilities of any machine. But the sophistication of computers and their algorithms is steadily increasing. Computer chess is a fine example. During the 1990s famous competitions were held between grandmasters and machines, such as Garry Kasparov against the IBM Deep Blue supercomputer. Chess is a complex game, and grandmasters think ahead not by calculating all the possible moves but by storing numerous images of previous configurations of the game. As recently as 2006, computers started to regularly win matches against the world’s best chess players. Part of this victory of machine over mind is due to the sophistication and advances in the chess software algorithms rather than the raw speed of the supercomputer. In 2010 a chess program running on a mobile phone reached grandmaster status!

Information pours into our eyeballs at an unmanageable rate, and the brain uses a variety of tricks to process the data as fast as possible. It is interesting how the the human mind does not usually process the the fact that I used the the word *the* twice each time in this sentence. Most brain tricks like this come from very low level subconscious preprocessing of visual information. What we perceive is a derivative of what we see. Our visual cortices discard what they deem unimportant and our brain fills in blanks with what we expect to see. You have little control over this as it is a basic function of your eyes and your brain. In this case, as your eyes are jumping from word to word as you read (this movement is called a saccade), it hits the word *the*, quickly registers it and jumps to the next word. Since the movement is a small one and the word *the* is something you see so often that your conscious mind essentially ignores it, your visual cortex figures that your eyes just landed on the same place as they were and instructs them to jump to the next chunk of letters.

As another example of how our brains process information: i cdnuolt blveiee taht I cluod aulaclty uesdnatnrd waht I was rdanieg. The phaonmneal pweor of the hmuan mnid, aoccdrnig to rscheearch at Cmabirgde Uinervtisy, it dseno't mtaetr in waht oerdr the ltteres in a wrod are, the olny iproamtnt tihng is taht the frsit and lsat ltteer be in the rghit pclae. The rset can be a taotl mses and you can sitll raed it whotuit a pboerlm. Tihs is bcuseae the huamn mnid deos not raed ervey lteter by istlef, but the wrod as a wlohe. Anazmig huh? Your brain is performing a complex process of pattern recognition: Not only is it recalling the words from memory given just the first and last letter of each word; it is also recalling phrases from memory and checking for the word that best fits the context.

The difference between a computer and a brain is that a computer performs a specific set of operations based on logic. A computer could be programmed to translate the above paragraph of mixed-up letters. You could tell it the rule—that everything but the first and last letters were mixed up—and it could perform a word search in an electronic dictionary, checking for unique matches. If there was more than one option, then it could look up and compare with previous sentences in which the words are used. However, you probably never saw the "rules" before, and your brain instantly managed to understand most of the paragraph. That is because your brain functions differently from a computer. We do not operate by carrying out sequential commands in a computer program; we think by forming an imaginary series of images and sounds recalled from our memories in parallel. We then automatically search our neural networks for similar stored memories. The brain searches for patterns, and the leading and end characters naturally distinguish the words the most. Moreover, we search for recognised words that make the most sense in the context of the preceding words. Finally, the most likely word is placed into our working short-term memory. The neural network seems to function in a probabilistic way, by recalling the most likely information from memory, and then piecing together a series of events that could happen by constructing sequences of imaginary images in our minds. Based on our memories of previous or similar events we decide the most likely action to take.

Intelligence and Brain Size

What makes us intelligent? The primary role of the brain is to enable survival by controlling behavioural patterns. In many insects and animals it seems that much of this behaviour is hardwired into the brain at birth and that behavioural patterns change very slowly. In more intelligent animals it seems that most of their behavioural patterns are learned after birth. This may be due to their larger brain capacities together with the ability of the brain to reconfigure itself as it acquires new knowledge. In "intelligent" animals the brain can think beyond day-to-day tasks and create images and thoughts of things that have never been perceived. I can picture in my mind a triangular wheel—something that does not exist—and I can also visualise how it would turn and how it would not work very well. I can reject a triangular wheel without needing to construct it to see how it functions. This ability seems to be present in some animals, but humans have developed it to an extent that has led to our current technologically advanced status.

The human brain differs from that of other animals in its detailed structure and in its size. The DNA sequences that hold our genetic makeup differ by just a few percent from a chimpanzee. That small difference combined with our larger brain capacities is the main feature that separates the two species. But why? The brain size of humans and our recent ancestors has steadily increased over the past three million years, although recent evidence suggests that our brain size has actually decreased by 10 percent over the past 20,000 years. Cognitive scientist David Geary suggests that this is a consequence of a complex society; individuals do not need to be as intelligent as previously to survive and reproduce.

Although it is sometimes said that we only use a fraction of our brains, this recent reduction in our brain size suggests that may not be true. Consider also that Inuit living in the polar regions have eyeballs that are 20 percent larger than those of equatorial dwelling people. This is because the amount of sunlight illuminating their world is lower at these high latitudes and their larger eyeballs compensate for it. They also have brains that are about 20 percent larger than the average human since the region that processes the visual information is larger. Inuit are no more intelligent than other humans—but it seems that their larger brains are necessary to process the visual data from a poorly illuminated world.

Whereas the physical size of a brain of a given individual does not correlate directly with his or her intelligence, it seems to be a requirement for maintaining the complex activities of a large body and its advanced behavioural characteristics. In *Descent of Man*, Darwin wrote, "No one, I presume, doubts that the large proportion which the size of a man's brain bears to his body, compared to the same proportion in the gorilla, or orang, is closely connected with his mental ability." We are certainly distinct from all other species when we compare brain and body sizes. The encephalisation index is a measure of the relative brain sizes of animals, defined as the ratio between the actual brain mass and its predicted value given the volume of the body it has to manage. A value of 1 indicates a brain capacity as expected for a body of that size. A cat has an index of 1 (similar to an elephant), rats

0.4, chimpanzees 2.5, and dolphins 5. Humans are at the top of this list with an encephalisation index of 7.5, meaning that our brains are 7.5 times larger than supposed necessary for a body of our size.

The brain requires a significant amount of resources to work. Even though it's just 2 percent of our body weight, when we are resting, 25 percent of our metabolism is used to provide energy for our brains. One of our closest relatives, the genus gorilla, requires 8 percent of its metabolic energy to function, whereas most mammals typically use just 3 percent of their energy for mental activity. It has been speculated that the growth in brain size in the *Homo* species is linked to dietary change as we became hunter-gatherer omnivores and learned to use fire. To maintain the energy requirements of our increasing brain size, we needed a protein- and calorie-rich diet. Whether or not dietary change is a cause or a requirement of our divergence in intelligence is a matter of debate. After all, many of the smartest animals eat vegetarian diets.

Counter-arguments against brain size being linked to the intelligence of a species can also be made. For example, the *Stegosaurus* dinosaur weighed the same as an elephant, yet its brain was the size of a walnut, 30 times smaller than that of an elephant. Even so, the *Stegosaurus* was as successful as the elephant in coordinating its massive body and allowing its species to survive for millions of years on our planet. A further example is some rare cases of hydrocephalus. About one in five hundred babies is affected by this defect, making it one of the most common disabilities. There are many causes, but ultimately it is due to an accumulation of cerebrospinal fluid in the brain. Most newborns die within months of being born, whereas those that live are usually mentally affected and have very low intelligence. However, in some cases people have been known to function normally with only a small fraction of the average brain. Their brains are reduced to a thin sheet of actual brain tissue spread over the inside surface of the skull, with cerebrospinal fluid filling the space usually occupied by brain tissue. Perhaps there is such a high level of computational redundancy in our brains and the processing power is so dense that even a reduced fraction of our neurons can be trained to accomplish many of the tasks that a normal brain can achieve. The brain has a remarkable ability to rewire itself and to forge new connections between neurons, not only as we learn but in the repair and reconstruction of damaged parts.

For some reason, a small number of people seem naturally gifted with the ability to perform astounding mental tasks, though this may just be the result of environmental stimulation. László Polgár is a Hungarian chess teacher and an expert in chess theory. Although he was not an outstanding chess player, he believed that "geniuses are made and not born" and wrote a book on the topic. After searching for a wife who could help him prove his theory, he married a school teacher and had three daughters whom he envisaged he could train to become exceptional chess players. Indeed, one of his daughters could beat him aged only five, and two become grandmasters.

Mnemonists are people with exceptional memories. The world record for remembering a string of random numbers is held by Lu Chao, who while a graduate student in China spent one year memorising the first 100,000 digits of pi. Pi is a transcendental number, a random string of digits that is infinitely long. After

24 hours of reciting numbers, Chao made a mistake after getting digit number 67,891 incorrect—a remarkable feat of memory! The evidence points towards this being a learned skill since the competitors in the world memory championships have all spent many years practicing and refining their memory strategies. These people can remember long strings of characters and entire books by memory, but often they cannot remember or identify faces any better than the average person. There are numerous examples of child prodigies who have extraordinary mental capabilities. Some people have no musical ability, no matter how much they practice, whereas some rare individuals have the ability to compose symphonies in their minds, "hearing" the musical parts of each instrument in the brain and writing down the notes as musical notation without actually performing or hearing the different instruments.

Child prodigies of today who stand out in society, in the arts or sciences, are influenced by environment. They rely on the knowledge gained through generations of creative individuals. Nurture clearly plays a role, but so does nature—our natural genetic inheritance that is passed on from our parents. Evidence for this is shown by the fact that identical twins separated early in life show closer IQs than random pairs of individuals. We are not born with knowledge of a number system, let alone the ability to multiply two numbers together. For most of us it is a learned skill, but some individuals find it very easy. Scans of the brain patterns of child prodigies reveal complex connected behaviour from across the brain.

Unfortunately, some of the people born with extraordinary skills have "savant syndrome" and suffer from developmental disorders. Famous examples include Kim Peek, who was portrayed in the film *Rain Man*. He was born with a damaged cerebellum in which the bundle of nerves connecting the two hemispheres of the brain was missing. He could read a book in an hour and remember its entire contents. He read two pages at a time, the left page with his left eye and the right page with his right eye! He could recall the contents by memory of over 10,000 books; however, his IQ of 87 was below the human average of 100. His brain was wired very differently from a normal brain.

Ants and Elephants

How do we compare to other living creatures? If we weighed all the living things on Earth—plants, mammals, insects, bacteria, fish and so on—we would find a total of about 500 billion tonnes. Most of this exists in plants and bacteria. If we discard the water content, the total human population would weigh about 100 million tonnes. Domestic animals would add up to 700 million tonnes and crops about 2 billion tonnes—an amount similar to all of the world's fish population. By weight, the biomass of ants and termites makes up one quarter of the total of all animals, or ten times that of humans. They occupy nearly every landmass and ecosystem on our planet from the dry hot deserts to the rainforests. They even thrive within our concrete cities.

Ants only have about 250,000 neurons compared to the 100 billion neurons of humans. Yet they are certainly one of the most successful creatures and not only by total weight and numbers. Ant colonies have social order; individual ants communicate with each other, and they can solve complex problems. They are one of a very few species outside of mammals that pass information on by direct tuition. Ants build very sophisticated underground cities that have highways, ventilation ducts, nurseries, farming areas and waste-disposal channels. Yes, that's right, ants are the only creatures besides humans that cultivate food, showing foresight and planning. Certain species of ants take the leaves of plants deep into their cities to special chambers to grow fungus on them. These chambers are kept at just the right conditions for the fungus to grow, which eventually supplies food for the ants. Ants also farm other insects for food, another trait they have in common with humans. They "domesticate" aphids, shepherding them to feed on plants and protecting them from other predators. The ants "milk" the aphids to extract a sweet liquid which is shared among the ant colony. These skills were developed 10 million years ago whereas humans began these traits just 10,000 years ago—if aliens had visited Earth prior to this time they may have thought that ants were the most intelligent species.

The moral behaviour of ants parallels that of humans. They can be altruistic and will sacrifice themselves for the benefit of the colony. They wage organised war against other colonies to capture territory and food resources. The victorious colony will often enslave their opponents. Other species of ants will settle their differences by choosing single individuals to fight one on one; the losing colony will submit and be slaves to the victor. Just a few years ago, a single ant supercolony containing a million queen ants was found to stretch from France to Greece. All the queens were related, and an ant which moved from one side of the colony to the other would be recognised as part of the colony and taken in.

I have only touched on some of the remarkable facts about ants to illustrate the power of collective behaviour despite each individual having a tiny brain. An isolated ant will not survive away from its colony; it will die since it can only function together with other ants. Each ant follows a set of basic rules and cannot accomplish a great deal on its own, not even fend for itself. Collectively, however, they appear as a remarkably intelligent superorganism.

What about the animals we perceive as the most intelligent? Dolphins and chimpanzees have about half as many neurons as humans. Elephants, in contrast, have 200 billion neurons, twice the number in the human brain. Elephants have been around for at least 30 million years. And although they have not evolved opposable thumbs, they have dexterous trunks. Elephants sustain their large brains on a diet of plants and fruit. They are self-aware: If you paint a mark on one side of the elephant's face and then show it a mirror, it will recognise itself and investigate the marks that are only visible in the mirror. Since its image is reversed in the mirror, the elephant knows its left from its right, which it must have learnt through reasoning, perhaps by studying its reflection in water. Elephants greet their friends, are happy to see them and grieve when they die.

How is it that ants are so successful and have developed sophisticated skills with such small brains? Why have they not evolved further or developed larger brains? Why did elephants not develop their skills to make and use fire? Are elephants just thinking more? We do not know the answers to these questions, yet. . .

As we interact with the world through our senses, we are "programming" our brains and filling our memory banks via the construction of a sophisticated neural network. The code for constructing our bodies is contained within our DNA, and every cell in our bodies has a copy of this information. Our "genetic code", the human genome, has been completely analysed. It consists of about three billion DNA base pairs, which corresponds to about a gigabyte of information. You can store your entire genetic code within the memory of your smart-phone. However, to store the information needed to map your neural network would take a hundred thousand times as much memory—your genetic code cannot hold enough information to fill your brain, which is why I argue that the brain begins its life mainly empty of knowledge and memories.

Brains have evolved over a period of several hundred million years to enable the finer control of the bodies of animals. Some species, and in particular humans, have learned how to use their brains beyond these day-to-day tasks, and our brain size has grown to accommodate this new role of advanced thinking. Operating a brain takes a significant fraction of our energy, which is ultimately derived from sunlight. We can contemplate surviving in a similar form as we are today as long as there are stars to provide energy and planets to provide the environment for our complex bodies. But what if the stars stopped shining, could we continue to survive in some form that was capable of conscious thought? If our brains operate in a way similar to our digital computers then I believe the answer is yes, and we can ultimately merge our minds with our machines to enable longevity on extremely long time-scales. While we develop the technology to accomplish this, we should first use our existing knowledge to explore and populate other worlds in our galaxy. This will ensure the continued survival of our species in the face of an ever more hostile environment on Earth.

Chapter 7
Life in Our Universe

B. Moore, *Elephants in Space*, Astronomers' Universe,
DOI 10.1007/978-3-319-05672-2_7,

After Berkeley I went to Seattle to start my second postdoctoral research position. The goal of my work for the next two years being to simulate the universe using a supercomputer. Seattle is the coffee and grunge capital of the world and my new colleagues laughed at my jar of instant coffee, but at least their music was good. Our enclosed suite of offices in the astronomy building had a wide range of uses at night, from multiplayer "Doom" tournaments to a climbing gym. The challenges ranged from who could hang the longest from their fingertips from the top of the narrow door frame, to circumnavigating the offices without touching the floor. Many late nights were spent in the underground College Inn pub next to the university, together with my good friend and colleague, Joachim Stadel. We played pool and darts and discussed topics from aliens and evolution to the origin of the Earth and the meaning of life. Today we are still working on some of those inspiring early ideas...

Surrounded by snowy mountains and steep volcanoes that Joachim and I loved to climb, we thought it appropriate to learn to snowboard, a much faster way down them than the usual hiking. We rented snowboards and set off after work to Stevens Pass, driving in my old beat-up American car that was like driving a tank crossed with a boat albeit with the addition of wheels. We took the chairlift to the very top of the mountain where the only means of descent was via a single steep black run—aptly named double-black diamond—the most difficult on the mountain. We had never snowboarded before but our strategy, formulated the night before at the pub, was clear and logical at the time of concept. We had concluded that if we made it down the hardest run, everything else would seem easy and we could progress quickly. Since there was no other way off the top of the mountain, we would not be able to back out and take an easier route down. We usually followed through on those ideas that somehow seem illogical when faced with the stark reality. It was snowing heavily as we walked over to the start of the descent and peered over the edge of a cliff. "Looks kind of steep, don't you think?" said Joachim. It appeared almost vertical. "No, I'm sure it's not so bad. But you go first," I replied—the snow was deep and soft so at the worst it would be a spectacular tumble and slide down to the bottom. Joachim was an expert skier, but snowboarding was new to him too. I wanted to watch his mistakes and hopefully avoid what I saw as an inevitable consequence of learning to snowboard the hard way. He set off and indeed immediately crashed and began the long but rapid descent to the bottom of the slope, upside down, on his back and head over heels. It was a spectacular show, like an out-of-control gymnastics display by an irregularly shaped snowball. I looked on scared at the top, knowing I had the same fate in store. The journey home was as memorable as the night's accomplishments. Our plan had worked. We had progressed well and were the last to leave the mountain. It snows a lot in the North Cascade mountains: Over the course of a year more than a hundred metres of snow can fall there, and that night was no exception. One problem with the beat-up old car was that the heater never seemed to work, and this night the windscreen wipers had given up too. It was below –15 degrees centigrade outside, and Joachim hung out of the window with a snow scraper in one hand and a can of antifreeze in the other to keep the windscreen defrosted and clear. The car made several 360 degree spins on the icy road while we careered on our way back to Seattle.

Life is fun, but dangerous!

Our galaxy appears devoid of life elsewhere, even though it may contain a billion stars with their own planetary systems possibly similar to our own solar system. As Carl Sagan wrote, "Absence of evidence is not evidence of absence." In fact, I would be shocked to learn that our solar system was the only place in the galaxy where intelligent life had developed. The Earth is already rather crowded, and it is in our nature to want to explore—indeed, there is no reason why our descendants should not spread throughout the galaxy to form a hyper-civilisation. In the next chapter I will explain how the Sun evolves in time, and although it has maintained conditions on Earth for over four billion years that are ideal for life to flourish, in a shorter timescale this will no longer be the case.

We have the necessary scientific and engineering knowledge to begin this exploration of our galaxy now. We could create a global population that would dwarf our tiny home called Earth, becoming something akin to how an ant or bee colony works: colonising the surrounding land by sending out teams of explorers until all the available space has been filled. Today we are concerned with our global ecosystem on Earth as we watch our energy requirements escalate and see the negative impact of our industrial activities. If we can survive this era and continue to progress as a species, within ten million years we could have filled the galaxy with life. In a further ten billion years we may be concerned about our global cosmic ecosystem as we watch our galactic energy reserves start to decline.

Life in the Solar System

Humans, like all life forms on the surface of the Earth, depend on the Sun for energy, air and food. Plants convert sunlight, carbon dioxide and water into organic carbon-based compounds such as sugars, producing oxygen as a by-product. The planet Earth is indeed a global ecosystem with a vast diversity of living things with their complex interdependencies. Sometimes, just the loss of one species can have a major impact across the planet over many levels of the food chain. Since life on Earth is so dependent on the Sun, we look for life elsewhere in similar environments, but perhaps life can also exist independent of a star. Life on Earth has adapted to very extreme conditions, some even existing on energy not derived from sunlight. Quite recently, "extremophilic" life forms have been discovered living in some of the most hostile environments on our planet.

The surface of the Earth is a thin crust of rock that is cracked in many places. The enormous broken pieces—the tectonic plates which roughly define our continents—jostle each other as they are pushed this way and that by the hot molten rock on which they float. As they slowly slip and slide past each other over timescales of millions of years, they cause earthquakes and create mountain ranges. The San Andreas Fault, which stretches through part of California, moves at a rate of about 3 centimetres a year, so that in about 20 million years San Francisco will be next to Los Angeles. Sometimes the plate boundaries lie deep under the ocean, such as the 10,000-kilometre-long Mid-Atlantic Ridge, where the North and South American plates move apart from the Eurasian and African plates. As they slowly separate, lava flows out of the crack, creating new subterranean land. Sometimes seawater flows into the cracks and re-emerges from hydrothermal vents superheated to temperatures much higher than the usual boiling point of water.

A fascinating discovery was made during the 1980s: Entire communities of life were found to exist around these vents at the bottom of the ocean, well away from sunlight, which does not penetrate to these depths. It is not just unicellular microbial life that can thrive within extreme environments. Two-metre-long bright-red wormlike creatures were found at the bottom of the ocean living alongside the hydrothermal vents. It was a mystery how these "tube worms" sustained themselves since they do not have mouths or digestive systems. The puzzle of how they survived was solved by Colleen Cavanaugh, who found that these creatures owe their existence to chemosynthesis. Their bodies are filled with billions of sulphur-oxidising bacteria which convert carbon dioxide, hydrogen sulphide and oxygen into organic matter, providing the giant worms' bodies directly with all the food that they need. It is a true symbiotic relationship—half of the worm's body weight is living bacteria!

The environment is extreme; thousands of metres below the surface of the sea the water temperature is usually just above freezing point, but around the deep ocean cracks it can be heated to above four hundred degrees centigrade. The pressure from the weight of the water is intense, three hundred times as high as we feel on the Earth's surface, which prevents the water from turning into steam. Many of the life forms in this environment are anaerobic, existing without oxygen, just like early life on Earth. The source of energy and elements for this process comes from the hydrothermal vents. Some types of bacteria even use the dim glow of the lava as the light energy that they need for photosynthesis. Another unique evolutionary strand recently found in this environment is the "scaly-foot gastropod", the only living creature known to have processed iron and pyrite into its skeletal structure. Its armour-plated foot has been investigated by the US military as part of their research into exoskeletal armour.

Perhaps most remarkable for their ability to survive the harshest of conditions are tardigrades, commonly known as water bears. They are microscopic water-dwelling eight-legged creatures about a millimetre long. When they walk, they swagger like bears, hence the name. Over one thousand species of tardigrades have been identified. They are found all over the world, from the Himalayan mountain tops over six thousand metres above sea level to a depth of four thousand metres under the surface of our oceans. Despite their name, these creatures can survive for at least 10 years without any water. They can live through temperatures as cold as −270 degrees centigrade or, at the other extreme, in water that is superheated to 150 degrees centigrade. In 2007 a colony of tardigrades was taken on a space satellite and exposed to the vacuum of space and the intense solar radiation from which we are protected by our atmosphere. After 10 days, on return to Earth, most had survived and laid eggs that hatched apparently normally.

Basic life forms can even survive deep within the Earth itself. In 2010 scientists exploring a South African gold mine two miles underground discovered an isolated self-sustaining bacterial community. Its energy derives from radioactivity. The bacteria depend on geologically produced sulphur and hydrogen for food—one of the few ecosystems found on Earth that does not depend on energy from the Sun. The hydrogen necessary for their survival originates from ancient water deposits and is liberated from the water by the radioactive decay of the elements of uranium, thorium and potassium in the surrounding rocks. These bacteria have been cut off from the surface of the Earth for millions of years. When they are exposed to our oxygen-rich atmosphere, they die.

Knowledge of such life forms that can withstand extreme conditions is very new. There are bound to be more discoveries over the coming years. It is incredible that life could evolve under the harsh early conditions on Earth with its constant bombardment by space rocks and fluctuating climate. It is also fascinating that life, even as we know it, can withstand a great range of conditions and can survive on very unique sources of energy. It is thus not so farfetched to think that bacteria and perhaps even tardigrades could travel through the solar system, immigrants on a space rock that was once ejected via a giant impact which sent debris into space. It is possible that life on Earth originated on Mars—we know that Mars had abundant water on its surface in the past and we know that some rocky pieces of Mars have landed on Earth. However, life is not going to make it to another star this way since the journey would take millions of years at the speeds at which asteroids travel and the chance of a fragment landing on another planet orbiting a distant star is negligible.

One of the biggest dangers to astronauts and life in space are cosmic rays—ultra-high-energy particles that fill our galaxy. They are charged particles, such as protons and helium nuclei, which are thought to gain their energies from inside the expanding shock fronts of supernovae remnants. When a star explodes, its debris expands into the interstellar medium close to the speed of light and carries a magnetic field which acts like a giant cosmic particle accelerator. The highest-energy cosmic rays are travelling very close to the speed of light and have energies that are up to ten million times higher than the protons inside the LHC beam tunnel! That is quite something: A single subatomic particle with an energy equivalent to the record-breaking 161-kilometre-per-hour cricket ball bowled by Shoaib Akhtar of Pakistan against England during the 2003 World Cup. If that single proton had collided with you, it might have knocked you over!

Cosmic rays fill our galaxy and are continuously striking the Earth at a rate of about one per square centimetre each second. Luckily, our atmosphere does a remarkably good job at shielding us from these particles—the cosmic rays collide with the molecules of air which prevent most from reaching the ground. Our atmosphere is a small coating on the surface of the planet—its density and pressure drop the further out one travels such that 99 percent of the air is below an altitude of 30 kilometres. When you fly in a commercial airplane at an altitude of 10,000 metres, the density of the air is one third that at sea level and the shielding effects from cosmic rays are much weaker. For each flight you take, you receive damage equivalent to what you would get from a chest x-ray, the same as a person on the ground receives over an entire year from the cosmic rays that penetrate our atmosphere. Cosmic rays are a serious threat to space travel since they not only damage biological life by destroying cells but also affect electronics by damaging their transistors and changing the charges stored in computer memory units.

There is another interesting location in our solar system that should be explored for new life forms—Enceladus, the beautiful blue/white satellite moon of Saturn. It measures only five hundred kilometres across. It should be frozen solid at this distance from the Sun, yet its interior is kept warm by a combination of heating from radioactive decays and from the gravitational field of Saturn, which continuously squeezes its moon. When the Cassini spacecraft passed by Saturn, it was manoeuvred to fly just 50 kilometres above Enceladus and captured images of enormous geysers of water erupting from the cracks on its icy surface. Moreover, organic compounds, such as methane and carbon dioxide, were detected in the water. Perhaps the interior oceans of Enceladus are filled with life, which would raise the fascinating question of whether its DNA structure is the same as life on Earth, or whether it evolved completely separately. If life has developed within its oceans, the only light will be a very dim glow of heat from the water which would radiate photons at radio frequencies. Creatures there may have evolved large radio antennae as eyes! We may even find life that is not carbon based. These would indeed be major discoveries.

Life Beyond Our Solar System

Life on Earth has evolved into numerous co-existing species: over 5,000 species of mammals; 10,000 of birds; 30,000 of fish; 1 million of insects and plants; and 10 million of bacteria. All of this happened within four billion years. The evolutionary steps from fish to apes took place in just the last half a billion years, whereas the steps from apes to humans took just three million years. It is the only example that we have of an evolutionary timescale, but it is an example. Half of the stars in our galaxy are older than 5 billion years, and some are 12 billion years old. Surely our galaxy must be filled with life at different stages of evolution. Just imagine for a moment how advanced life could be if it developed for a billion years beyond us. The capabilities of a future advanced and highly evolved civilisation, be it our own, or alien life on a distant planet, are probably beyond our imagination.

In 1995 the Swiss astronomer Michel Mayor discovered a planet orbiting the solar-like star 51 Pegasi which lies many light years from the Sun. Astronomers are now routinely finding and characterising these so called extra-solar planetary systems, many stars even having multiple planets orbiting them. It is hard to see the planets directly since they are so much fainter than the stars around which they orbit, but we can detect them and measure their properties in several ways. Just as a planet orbits the Sun because of gravity, the Sun also moves in a regular periodic way owing to the gravitational attraction of its planets. These small wobbles of stars can be observed, and the period and strength can be used to determine the mass of the planet and to measure how far it is from the star it orbits. Another planetary detection technique uses the fact that the brightness of the star decreases if a planet passes directly in front of it in our line of sight. By monitoring the change in brightness over many days, we can characterise the orbit of the planet and measure its size. The Kepler space telescope has been doing just this. It has discovered numerous new extra-solar planets at a rate consistent with most stars hosting planets in our galaxy. Detecting Earth-like planets, characterising their atmospheres and compositions to detect the bio-signatures of existing life is one of the most fascinating research topics today.

An "Earth-like" habitable planet broadly describes a mostly rocky planet, large enough to hold an atmosphere, close enough to its star so as not to be continually frozen but not too close such that water would boil from its surface. The other class of planets is the larger "gas giants" like Saturn and Jupiter, which are mainly hydrogen. These are in a sense failed stars since they never became massive enough for their internal temperatures to ignite the process of nuclear fusion. Our computer simulations showed that habitable Earth-like planets should be very common.

Understanding how life emerged is a fascinating problem to study, and in principle easier than trying to understand what happened during the first millionth of a second of our universe. It is easier to answer because life has evolved according to laws of physics and rules of chemistry that we know and understand very well. The instant after time began, our known physics fails. So we are in a much worse position to achieve an understanding of the origin of our universe. The "primordial mud soup" from which life crawled is very different from the "primordial fireball soup" of the early universe. Because we have good ideas as to how life developed, they can be tested through laboratory experiments and computer simulations.

Water on Earth is filled with molecules. They move around bumping this way and that, colliding with each other and sometimes sticking together because of the electromagnetic force. The tendency for molecules to organise themselves into complex configurations is the starting point for the origin of life. Perhaps the first reproductive molecules formed near the oceans' volcanic vents, which in turn led to complex structures such as ribonucleic acid. These were followed by prokaryotic single-celled life that underwent generations of mutations and natural selection until we reached the stage of computation—the brain. This is indeed an amazing series of events. And although we do not understand these early steps in great detail, they are certainly possible given the timescales, boundaries and behaviour of the laws of nature.

Life managed to evolve on Earth. There is no reason why it should not have occurred elsewhere in our galaxy. If just one of those distant planets developed life akin to humans, with the desire or need to leave their planet, they could already have filled the galaxy with life.

We have the technology today to begin to design and construct spaceships that could take a breeding population of humans to nearby stars. It represents a major project that would take the cooperation of many countries from across the world. The financial cost is huge but less than the global expenditure on war. Within just a few hundred years we would be ready to begin the ultimate era of human exploration—that of our entire galaxy. Let us send several pioneering craft in different directions and to the closest stars that we will have already determined to have habitable planets with atmospheres not too unlike our own.

On arrival they could begin building a new civilisation. They would not have to reinvent everything; they could take their knowledge and tools with them to quickly establish their societies and cities. If each spaceship were to take one hundred pairs of humans, it would only take about 15 generations to reach a million people on their new planets. And they would begin their colonisation with all the knowledge about science and medicine that we now have. As these colonies develop, they can still communicate with Earth, learning about new developments as they grow. It would not be like a video conference though, since it would take 10 years for a message to be exchanged with the closest stars. The Earth would broadcast a continuous stream of information to the new colony. Within a thousand years these first explorers would have built up the resources and infrastructure to each construct and send several new spacecraft in new directions, to continue the process of populating the galaxy.

A wave of colonies would spread outwards from Earth and begin to fill the galaxy with life at a rate which would depend on the speed of the spaceships that we could construct. The typical distance between the stars is about five light years, and the galaxy is two hundred thousand light years across. If we could build spacecraft that could travel at 10 percent of the speed of light, the average journey between stars would take 50 years, which is within the lifetime of the astronauts. Adding in the time to colonise each planet, we could literally populate the galaxy within ten million years. That seems like a very long time, and it is for our short lives. But to a species, it is just better than average survival rates. A million years is a small period of time in terms of geological and evolutionary events, let alone the enormous cosmological timescale—billions of years—in which stars live. On our 24-hour universal day the project would be accomplished within one minute after midnight.

What is the reward for such an effort? Imagine our descendants witnessing the landing of these spacecraft, the first footsteps on new worlds. When Apollo astronauts landed on the Moon, the world watched transfixed, proud of the achievement. The entire planet felt part of a united purpose and could do so again. As we optimise and design these giant craft, with their self-sustaining ecosystems, we will advance the state of knowledge on Earth. The spin-offs of such an investment cannot be predicted in advance; perhaps we would learn to control nuclear fusion as a means of flight energy, thus giving our planet a limitless resource of clean energy. Eventually, if our species wishes to continue to exist, we will have to leave the planet—the climate on Earth has been suitable for life to develop for billions of years, yet even the Earth will not last forever. Already, 80 percent of its life span as a habitable planet has passed. As we shall see later, the conditions on Earth are slowly becoming less hospitable for the hosting of life.

The Fermi Paradox

If we could begin to colonise the galaxy now, and it would take less than ten million years to complete, could it have already happened? If so, why is the galaxy not obviously full of superintelligent life? Humans did not arrive here in spaceships; we clearly evolved on this planet. Life today is genetically linked to the life that existed on Earth a million years ago. We know this through the study of plants and insects preserved in ancient ice at the bottom of glaciers. Of course, it is possible that the bacteria-like seeds of life were transported here by an advanced civilisation several billion years ago—their means of transport no longer visible, buried like a fossil deep inside rocky strata.

Why are there no signs of life in our galaxy? This is a paradox that has puzzled many people, most notably the physicist Enrico Fermi in 1950. There is no compelling evidence that alien life forms have ever visited us, nor do we see any sign of other life as we peer out into the cosmos.

In 1959 American astronomer Frank Drake wrote his famous equation describing the probability of life existing elsewhere in our galaxy. A year later he searched for alien signals by observing two nearby stars in the radio wave part of the electromagnetic spectrum. Various SETI (Search for Extra Terrestrial Intelligence) projects are taking place using some of the world's large telescopes to monitor many of neighbouring stars. The projects look for signals deemed "intelligent" by searching through the electromagnetic spectrum, looking for non-random complex behaviour, in the same way you search for a desired radio station. Turning the tuning dial scans the small part of the electromagnetic spectrum in the radio wavelengths (spanning about 10 centimetres–10 metres), amplifying the signals from stations.

The lowest-energy photons we can measure are very long radio waves that penetrate through the ocean and are used to communicate with submarines—wavelengths of three thousand kilometres which are broadcast from two large ground stations, one Russian and the other American. The submarines can receive the signals but cannot transmit anything back since the transmission antennae are many tens of kilometres across. The highest-energy photons are gamma-ray photons, which can arise from nuclear fusion or radioactive decay. They have wavelengths less than a few picometres, which is smaller than the size of an atom. So far, the SETI searches have revealed nothing.

Intelligent beings living on distant planets could also be looking for signs of life among their neighbours. If they were just a little more advanced than us, they could have known for over a billion years that our own planet most likely hosted some forms of life. The signatures of life on Earth are present in the abundance of numerous organic molecules in our atmosphere, which could be imaged by a remote alien species and analysed for detailed composition. For two billion years of our history on Earth, this was all the evidence that could be learned from far away. In the last 80 years we began broadcasting radio and TV signals in the radio and microwave bands. These signals leave Earth in all directions and have already filled a sphere with a radius of 80 light years. Within this distance are over one thousand stars that have received our broadcasts and aliens may have already dispatched a high-speed spacecraft to pay us a visit. If they are advanced enough to use nuclear or antimatter propulsion, they could reach the Earth anytime soon. Let us hope they are friendly and, after watching us develop our primitive nuclear weapons, have not come to wipe us out before we become more advanced. Our electromagnetic signals spread outwards into the galaxy at the speed of light. In 1,000 years, potential knowledge of our presence will have reached over 100 million stars.

In 1974 we not only listened for messages, but we broadcast an extremely powerful 3-minute radio message towards a star cluster that contains one million stars 25,000 light years away. It was transmitted using the giant three-hundred-metre Arecibo radio telescope that is usually used to map the structure of our galaxy. The encoded message contained some of our basic knowledge about atoms, DNA, a map of the solar system and more. We expect no reply for another 50,000 years though, which is how long it would take before a message could come back from any remote listeners.

If humans can survive their selfish tendencies to destroy each other and thus themselves, they could evolve inhibited only by the laws of physics for millions of years. We would have the technology to create giant beacons in space, beaming out our messages, easily visible across our Milky Way and reaching even to the most distant galaxies. Perhaps intelligent life elsewhere does not want to be found. Or perhaps our alien-filled galaxy does not broadcast in the electromagnetic spectrum. Perhaps they have discovered a different and more efficient way of communicating, such as with gravitational waves, neutrinos or some physical process that we have yet to discover.

There are many other logical solutions to the Fermi paradox:

Perhaps there is no other life out there. It just did not happen elsewhere because the conditions were not exactly right. The chances of reproductive organisms forming from the random motions of interacting colliding molecules may be incredibly small. That would make me feel very lonely in this vast universe.

There may be abundant life, but nothing comparable to humans. After all, there were millions of species on our planet prior to *Homo sapiens* and none of them have evolved beyond using a few primitive tools, let alone having mastered the use of fire. Perhaps it is the same on other planets that orbit distant stars; some may be teeming with life which we would call primitive.

Even if intelligent life were to develop elsewhere, it would seem to be an intrinsic part of evolution that certain individuals of a species seek to survive and profit at the detriment of the welfare of the global society. Thus any intelligent society will self-destruct before it reaches the global coordination necessary to begin exploring the stars.

There is the possibility that, when life becomes intelligent enough, it creates a machine that can "think"—a computational mind comparable to our own. This may inevitably lead to the rapid demise of a species that has evolved through the long process of natural selection in the presence of machines that can undergo exponentially fast evolution through intelligent design.

I will discuss some of these interesting possibilities in more detail in the final chapter. But I prefer the notion that there is life, plenty of it and some of which is indeed far more intelligent than us. We just have not seen it because we are not looking in the right way or in the right places. If we were to find life on a distant planet, it should not make us feel less insignificant in our universe, but it should fill us with a desire to understand far more. Perhaps another life form has already figured out the answer to my question regarding the first instant in time: Why and how did the universe appear?

A Journey to the Stars and Beyond

The two Voyager spacecraft were launched in 1977 by NASA with the main purpose of exploring the planets in the outer solar system. They still transmit data back to Earth using instruments powered by small nuclear reactors that have a lifetime of about 50 years. In 2020 they will leave the confines of our solar system, the region that is still influenced by the Sun's wind of particles that are constantly escaping from its surface, and will travel through the vast empty interstellar medium of our galaxy. They are our most distant man-made objects, already over 10 billion kilometres from Earth. Voyager 1 is travelling at 17 kilometres per second. In about 40,000 years it will be close to a star named AC+79 3888, close to Polaris in the constellation of Camelopardalis.

As a statement regarding the intelligence of the human race, NASA placed records on the spacecraft which contain a short summary of our knowledge and culture. These are actual phonographic records made of gold-plated copper encased in aluminium. They contain a selection of Earth music from genres as diverse as Chuck Berry and Beethoven. This was the medium of choice for music for most of the last century until magnetic tapes took over, then CDs and so on. The records also contain images encoded in analogue form composed of 512 vertical lines, which contain key findings from mathematics, physics and other scientific disciplines.

It is unlikely that these craft will ever be discovered. Yet, although the onboard power is expected to cease to function around the year 2025, the Voyagers and their contents will most likely travel for billions of years through our galaxy unharmed. They are symbols of the state of development of our species that an advanced alien civilisation could easily decode. They could then recognise that the inhabitants of our small blue planet are relatively harmless and primitive, and perhaps like Darwin, they will listen to our music and look at our behaviour and call us "savages".

Within 20 light years of Earth there are about 80 visible stars. Space-based observational missions that are being proposed and designed now would allow us to determine not only whether they have planets but where the planets orbit and what their atmospheres are composed of. We will know where to go, and we already have the resources to design and construct a spaceship capable of sustaining the one or two generations of people that are necessary to make this journey. I suspect there will be no shortage of volunteers to depart on the first one-way trip.

A commercial airplane has a top speed of about 1,000 kilometres per hour and would take 15 days to reach the Moon, 6 years to get to Mars and a further 137 years to reach Saturn. At this rate it is going to take a long time to reach the nearest star, about five million years in fact, so we need something a little faster. Our fastest existing spacecraft is Voyager 1, travelling at 60,000 kilometres per hour. Like fireworks, it obtained part of its speed from burning combustible fuel; the reaction of the exhaust blowing out of the back of the engines causes the spacecraft to accelerate faster forwards (this is a consequence of Newton's third law, which states that every action has an equal and opposite reaction). Once the fuel is gone, the spacecraft will just continue to travel at the same speed forever (a consequence of Newton's first law of motion).

Travelling with a rocket-propelled spacecraft at Voyager 1 speeds will be our low-budget mission to the stars, but later we will see what we are really capable of. We would reach the Moon in just a few hours. This is the farthest that humans have so far explored in person. Outside the spacecraft is the near vacuum of space. On Earth our atmosphere is filled with oxygen, nitrogen and other elements: Over a billion billion particles are packed together in a single cubic centimetre of air. The Moon has no atmosphere; its gravitational field is not strong enough to prevent the atoms from escaping its surface all together. With no atmosphere there is no natural erosion from wind and rain. The footsteps of the Apollo astronauts from 50 years ago are still imprinted on the Moon's dusty surface as if they were made yesterday. They will remain preserved for thousands or even millions of years, until eventually another meteoroid or asteroid strikes the surface and completely erases the evidence of our presence.

A cubic centimetre of space beyond the Earth contains just a few particles. We hear no sound—there is no medium through which a sound wave can propagate, at least on the scale of our human ear. On Earth we are kept warm by the random collisions of numerous molecules of air with our skin. Those molecules are kept moving by the radiation of the Sun. It is extremely cold in space. If you happen to find yourself outside the spacecraft, your skin would radiate heat until it cooled down to the temperature of the photons hitting your body. The side of your body away from the Sun would cool down to a temperature close to absolute zero. Away from our light-polluted atmosphere, the sky glows bright with stars of the Milky Way. They do not twinkle as we see them from Earth, where the light is constantly navigating different paths through our turbulent atmosphere. Away from the womb of Earth, where we evolved and where all of human history has taken place, most of our senses are useless. We cannot hear, feel or smell anything.

Our last excitement for a long period of time will be passing by the cold and barren icy surface of Pluto, 10 years after departure. The Sun's light is feeble at this distance and warms its surface to a temperature of only −230 degrees centigrade. Communication with Earth is already difficult and verbal conversations are not possible because it takes the signals about 6 hours to travel in one direction. It is a soul-destroying sight on our journey to see the Earth fading from a pale blue dot to nothing as we leave the farthest reaches of our solar system. Our Sun eventually becomes just another distant star in the night sky.

Time ticks on, year after year, century after century, millennium after millennium as we journey across the empty expanses of our galaxy. After 80,000 years we finally approach the next-closest stars to Earth, Alpha Centauri, 4.4 light years away—a binary system of two stars, each of which is similar in mass, size and age to the Sun. The two stars in Alpha Centauri move on elliptical orbits around each other that take them between 1.7 and 5.3 billion kilometres apart every 40 years. Computer simulations have shown that it is possible that an Earth-like planet could exist in the habitable zone around either star orbiting at a distance of about 100 million kilometres from one of the stars. In 2012, an Earth sized planet was discovered orbiting Alpha Centauri B. The view of its two suns must be quite spectacular!

At this speed the time taken to reach the nearest star roughly equals the amount of time required for most of the recorded intellectual history of the human race to unfurl. During this time, thousands of generations of people will have been born, lived and died on the craft. This is too long a journey, unless suspended animation or cryostasis becomes a reality. The principle is simple. If a body is cooled down to about −200 degrees centigrade, all the essential biological activity stops. Everything is frozen in time and can be thawed out at some arbitrary point in the future. It doesn't quite work yet though. If you simply freeze a human, the cell structures are destroyed by the ice crystals that form around them. Researchers working on cryopreservation are looking at ways of soaking bodies in a type of antifreeze, like ethylene glycol, which soaks into the cells and prevents them from crystallising.

Despite the failure to resuscitate any human who has been frozen for a long time, it is possible to pay to be placed in suspended animation before death. The selling point of the companies that provide this service is that in the distant future, the technology will have advanced such that our descendants will be able to bring you back to life—despite the fact that they may not want to! We know that the technique works for simple parts of the human body, such as blood or embryos. We just have not mastered it for the body in its entirety. Many insects, fish and amphibians naturally produce cryoprotectant chemicals in their bodies to minimise damage from freezing. Those tough little water bears can survive extreme temperatures because they have very little water content and are filled with the glucose sugar trehalose, which maintains cell structures in a gel-like state even when frozen to −270 degrees centigrade.

I imagine that one day it will be possible to place humans into a low-power hibernation mode for long periods of time. Today it is only an emerging future technology. We want to see more immediate results, so let us look at ways of speeding up our journey using existing technology. Is there a way of making the journey to the nearest stars within one human lifetime? This would require a spacecraft that could travel at least 30,000 kilometres per second—that is 10 percent of the speed of light and over a thousand times faster than our rocket-propelled ships are capable of. The surprising answer is yes, we could start to design and construct such a spacecraft today. There are two known ways: laser powered sail crafts and nuclear pulse propulsion.

The principle of the first method is that photons of light have a tiny momentum which is imparted to any object with which they collide. In the early universe, the photon pressure drove the expansion of space. In dying stars, the photon pressure pushes away the outer layers of material. In the solar system, the radiation from the Sun strikes the icy surface of comets and creates long dust tails as it literally blows away the material. This is why the tails of comets always point away from the Sun. Our spacecraft would have a 10- to 100-kilometre-diameter very thin metallic "sail" which would need to be constructed or unfolded in space. Then, extremely powerful gigawatt lasers on Earth, or orbiting the Sun, would focus their light beams on the sail, pushing and constantly accelerating the craft all the way to its final destination. They would have to operate continuously for many years. And although this technology uses "clean energy", it would take a large fraction of the energy output of our planet to provide the propulsion.

The photon pressure is so small that the craft barely moves at first. But it begins to accelerate faster and faster until, after several years, it will have reached its top speed, which is over 10 percent that of light. Slowing down the craft when it reaches its destination would be difficult, but the sail could be detached and sent ahead of the spacecraft. The laser beams from the solar system would be used to reflect back onto the craft, slowing it down over the final part of its journey. The first space mission to use a "solar sail", which captures the photon pressure from the Sun, is the ongoing Japanese IKAROS (Interplanetary Kite-craft Accelerated by Radiation Of the Sun) mission, launched in 2010. We could develop the technology for an interstellar mission, but the cost of this method is very high.

Perhaps the method of choice for propelling ourselves to the nearest stars is to use the raw energy contained in matter and to detonate a series of small nuclear fission or fusion bombs immediately behind a specially designed spacecraft. A huge steel or lead shock-absorbing plate would take the impact of the bomb debris, which occurs in a few nanoseconds, spreading the impulse over a few seconds. The spacecraft would literally be pushed forward by the blast wave of particles coming from the nuclear explosion. After about a thousand blasts, the spacecraft would be accelerated up to 10 percent of the speed of light. Allowing for speedup and slowdown, the journey to the next star could take as little as 50 years; it could thus be made in a single human lifetime.

A design for this mission was already created in 1950 at Los Alamos National Laboratory in New Mexico, named Project Orion. British physicist Freeman Dyson worked on the project and designed missions of several sizes, up to the Super Orion craft, a 400-metre vessel that could hold a small city. However, interest in nuclear pulse propulsion declined in 1963 following the Nuclear Test Ban Treaty signed by the United States, the Soviet Union and the United Kingdom. This agreement banned nuclear weapons tests in the atmosphere, underwater and in outer space.

Given the limitations of the speed of light, and the fact that warp drives from *Star Trek* are in the realm of science fiction, the ultimate mode of rocket propulsion known to be possible uses antimatter as fuel for propulsion. Protons and electrons have antipartner particles that have opposite charge and opposite magnetic properties. The existence of the antiproton was predicted by the British physicist Paul Dirac in 1933 and was discovered experimentally in 1955. When a proton comes into contact with an antiproton, the pair annihilate each other and their entire rest masses are converted into pure energy. It is the ultimate form of energy generation. Unfortunately, it is very difficult to manufacture and store antimatter using today's technology. It is the world's most expensive substance to produce, costing 50 trillion euros per gram. But the technology can certainly be improved to bring the manufacturing costs down considerably.

Antimatter drives could potentially yield enough energy to send spacecraft travelling at close to the speed of light. The journey between stars could be made within a decade and with the bonus that, at these speeds, the traveller receives the benefits of time dilation. That is a consequence of Einstein's relativity that I mentioned earlier. The faster you travel, the slower your clocks run and the slower your body ages. At 99 percent of the speed of light, time is stretched by a factor of seven. To someone watching from Earth the journey might take 10 years, but to the astronauts it would take only a year and a half!

It would take our nuclear-powered spacecraft about one million years to cross the entire galaxy. It would be a spectacular journey, during most of which time you would have to be cryogenically preserved, but could be woken at key moments to see some of the spectacular sites on the way. In principle, with future technology using antimatter drives, we could continuously accelerate our spacecraft to speeds closer and closer to the speed of light, allowing us to cross the entire galaxy within a human lifetime. But at these speeds we have the additional complexity of preventing the spacecraft being destroyed from the collisions with large molecules and dust within the interstellar medium.

After we have travelled through our galaxy, what next? What about the galaxies beyond our own? Perhaps many are already filled with life. The nearest large galaxy to us is Andromeda, which is 2.5 million light years away. That would indeed be a long and lonely journey to make. The distance to the edge of our visible universe is called the horizon. This is the largest distance we can ever measure or travel to, since it is the distance that light has travelled since the beginning of our universe. We cannot see past this distance: Light from farther away has yet to reach us. There is no way of communicating information on scales larger than our horizon, which is why we can never measure the true size of our universe. Our horizon distance grows with time because, as the universe gets older, the light from more distant regions has had more time to reach us. Today, our visible universe is a vast 90 billion light years across. But like atomic matter, our universe is mostly empty space.

If we spread out all the stars from all the galaxies within our horizon, there would only be one star in every 10 billion cubic light years of space. The universe is mostly empty space between the galaxies that we see. If we travelled in a straight line, we could cross the entire universe without colliding with a single star. In fact, you would have to travel across the universe a trillion times before you had a good chance of directly colliding with one star. This is why the night sky is dark and not glowing brightly with light from an infinite number of distant stars. In most directions there is simply nothing.

Even if our galaxy were empty of life, I hope that I have convinced you that we have the capabilities to fill it with our own species in the near future. But what about the far future and the question of how long life could exist? The gas supply for forming new stars is slowly being depleted, and more and more stars are beginning to fade and die. The energy sources on which life depends today will begin to dwindle. This would be a good place to begin to discuss the future of our universe and the prospects for our species to continue given all the theoretical and observational knowledge that we have. This is particularly timely given the cosmic matter and energy inventory that we have established. Can life exist for eternity, becoming ever more advanced and sophisticated? Or is life itself just a fleeting event in cosmic history viewed from the eyes of our ancient universe?

Chapter 8
Tomorrow

B. Moore, *Elephants in Space*, Astronomers' Universe,
DOI 10.1007/978-3-319-05672-2_8, © Springer International Publishing Switzerland 2014

We were headed for Hueco Tanks in El Paso County, Texas, a legendary climbing destination. It is an outcrop of weathered syenite granite strewn with pockets and holes from which you can hang upside down—it was like climbing in a perfect rock gym but with stunning scenery. The caves amid the rocks had been home to the Paleo-Indians, the first settlers who entered the Americas across the Bering Strait from Asia during the middle of the final glacial period 20,000 years ago. The weathered holes in the rock collected abundant water, and the cave walls provided the canvas for some of the most beautiful Mogollon and Apache Indian paintings. Joachim and I stood at the bottom of the spectacular Sea of Holes, one of the most famous routes in the park. We were looking for the first bolt, the small metal clip drilled into the rock and to which the lead climber can attach the rope, thus giving protection and lessening the danger of a potential fall. It was 20 metres above the boulder-strewn ground on a vertical cliff, and a fall before that point would not be good. We debated over who should lead the scary climb, each hoping to give the honour to the other until finally I managed to persuade Joachim to set off. At the midway point up the 100-metre vertical climb there is a cavelike hole large enough to fit two people: a good place to rest and enjoy lunch. Our legs dangling over the void below, we stared across the desert, the solitude broken only by distant explosions at White Sands military base and nuclear bomb testing facility. "If only we had a couple of beers," said Joachim. I smiled and suggested that he check his backpack. Now, you should understand that Joachim would go to the extent of shaving the labels off his climbing shoes to save weight in his quest to ascend sheer featureless vertical slabs of rock. He reached into his backpack and found the two cans I had stuffed in earlier without him seeing. Luckily, he saw the funny side and handed one to me. At the end of the day we trudged back to the car park, dusty and dirty, with arms which felt like lead but happy about our achievements. I sat in the evening sun while Joachim went to ditch all the day's trash in our rental car, a white one like most other rental cars. I watched in amusement while he opened the back door of a car that was indeed white and sort of resembled ours. He reached in and pulled out a backpack which was sort of similar to his own and started to take things out—some frilly knickers, a bra.... He looked a bit puzzled but stuffed them back in and added the day's trash on top. When he came back over, I explained that he did not have any girl's underwear in his backpack, that it was not actually his backpack and it was not even our car!

To present the history of our universe and the development of life on a timescale that we can more easily comprehend, I rescaled and compressed 13.8 billion years into one 24-hour day. In this chapter I would like to discuss how our galaxy evolves and changes over the next period of time that is equal to twice its current age—during which time the Sun stops shining and the Earth is incinerated and shredded to dust. I will discuss the possibility that life can survive and continue for 13.8 billion years into the future—to the end of tomorrow on the same timescale. It is a journey into the future of science fiction proportions, but one that is ultimately possible and, of course, necessary if we wish to survive as a species. However, our two most imminent threats in the first second after midnight are the ecological disaster that could result from our industrial activities, which are causing a runaway amplification of the greenhouse effect, and the possibility of a global war culminating in the nuclear destruction of the Earth's surface.

President Franklin D. Roosevelt was sent a letter from Einstein in 1939 which began by stating that a new form of energy could be obtained from the nuclear fission of uranium. But he also warned that it might be possible to construct a new form of devastating bomb from this material. Already by 1945, scientists had mastered the techniques needed to perform the world's first controlled nuclear explosion. The principle behind the fission bomb is to put together enough of the unstable element uranium-235 so that it will undergo a chain reaction whereby its nuclei literally split apart, giving forth a fraction of their rest mass as energy. It is rather like the reverse of nuclear fusion.

The process starts when an unstable uranium-235 atom collides with a free neutron that triggers its decay into two new atoms, such as caesium and barium. These fission products are themselves radioactively unstable and are the "nuclear fallout" from the explosion. The decay also emits three new neutrons and some high-energy gamma-ray photons. The 3 neutrons can trigger the same process in 3 new atoms, resulting in 9 neutrons, then 27 and then 81. A runaway decay process ensues, resulting in an explosion of nuclear energy if a critical mass (about 50 kilograms) of the substance is put together. You could literally carry a nuclear fission bomb in two 25-kilogram suitcases—just don't put them next to each other. Nuclear power stations manage the reaction using silver control rods that they can lower into the uranium, absorbing the neutrons and slowing down the reactions. It is when those control rods fail that the runaway reactions lead to a meltdown and possible massive explosion.

In the 20 years following the US development and deployment of nuclear bombs, the Soviet Union, France, China and the United Kingdom all followed and stockpiled nuclear arsenals. These countries signed a "friendly" non-proliferation treaty in 1968, but since then, India, Israel, Pakistan and North Korea also constructed and tested nuclear weapons but have not signed the international treaty. At the peak of the cold war in 1988 there were over 60,000 nuclear warheads. Today, after the agreement to slowly dismantle the nuclear arsenals, there are about 10,000 actively maintained nuclear weapons and another 20,000 stockpiled.

The destructive capacity of existing nuclear weapons is sufficient to destroy all the world's major cities several times over. Climate modellers predict that a nuclear war is likely to have the same global consequences as a giant asteroid strike, which would change the planet's climate sufficiently to wipe out many of the existing species. While we seem to have been content in the past decade or two that the Earth's superpower countries have been friendly, it should not be forgotten how quickly this situation can change. In the last century alone, 200 million people are estimated to have died by human decision during over 300 separate wars. The pace at which new conflicts are emerging seems to be no slower than the previous century, and the possibility of another global world war remains. It is a strange facet of human nature that the human race can be inspired by certain individuals and can follow their orders and ideas seemingly without question. As Adolf Hitler said, "How fortunate for leaders that men do not think." This may be an inherent trait of life that emerges through natural selection, and one over which we have little control.

24 hours + 00:05:00

Giant Impacts

While we scramble today to avoid man-made catastrophes, we must also plan and prepare for the equally devastating possibility of a giant asteroid impacting on our Earth. The periodic decimation of species on Earth is thought to be due to random collisions between the Earth and asteroids that are several kilometres across or larger. From the fossil record we can estimate the number of species that lived within different time periods. There were five events over the past 300 million years during which about half the existing species became extinct. If giant asteroids were the cause, then they occur about every 50 million years. The last one was 65 million years ago, so our planet is already overdue for another. The interval appears to be more regular than expected from random events, which has been attributed to the passage of the Sun through the galactic spiral arms where the density of stars is higher. These spiral patterns of many galaxies closely resemble the appearance of clouds within a tropical hurricane, although the physics that leads to the patterns is very different. Hurricanes are due to the complex turbulent flow of a gas on the surface of a rotating sphere; spiral patterns in galaxies result from complex gravitational interactions between the thin disc of rotating stars. If the density of stars in the neighbourhood of our Sun increases, another star may pass close enough to the outer solar system so that its gravity perturbs the orbits of distant comets and asteroids that currently orbit at a safe distance from Earth. This could send a shower of kilometre-sized icy boulders hurtling towards the inner solar system where Earth orbits.

Most of the stars you can see are rotating around the galaxy with the Sun, although they all have additional random motions of some tens of kilometres per second such that they will change their positions over time. However, the constellations have barely changed since sketched by the Babylonian astronomers around 1000 BC. It would take about 50,000 years for the stars to move enough such that familiar constellations do not exist anymore. The Big Dipper will eventually lose its handle and flatten out. That is a good fraction of the age of human history. Our distant ancestors gazed upon a very different night sky. It takes an even longer timescale to see geological changes on Earth. The continents move slowly, carried on the molten rock that swirls around beneath them. Yet in 50 million years Africa will have nudged upwards, joining completely with Europe and eliminating the Mediterranean; Australia will collide with Asia, and California will be next to Alaska.

The Earth travels around the Sun at a speed of about 30 kilometres per second, and any leftover rocky debris from the formation of the planets impacts the planet at about this speed. Our atmosphere prevents the smallest meteoroids from reaching the Earth's surface due to the friction against the air molecules, which rapidly heats the rocks and causes them to glow brightly—we see these as shooting stars. These are typically centimetre- to metre-sized objects that are mostly burnt into dust.

About 20,000 centimetre-sized objects strike the Earth every day. They burn up in the air and break up into fragments before landing. Larger objects are more rare. Every day a rock about half-metre across impacts our atmosphere from space, and each year a 5- to 10-metre object will hit the Earth, somewhere.

You can find meteoritic dust lying on the ground outside using a powerful rare-earth neodymium magnet. Look at the dust on the magnet—you might be able to see tiny pieces of iron, made spherical as they became molten because of the friction with the air as they fell into our atmosphere. These will not hurt you. In fact, you have probably been hit by plenty of them already without noticing. It is the asteroids larger than a football stadium that could cause significant destruction.

One of the last significant impacts recorded on Earth was the Tunguska event in Siberia in 1908. No one lived very close to the impact site, but 100 kilometres away people reported seeing a burning object as bright as the Sun falling to Earth. It was followed by an enormous explosion that shattered their windows; the resulting atmospheric shock wave was detected as far away as England. In 1927 the first explorers reached the area, motivated to find the iron remains of the meteorite to aid the Russian industrial expansion. Instead of an impact crater they found that a vast region of forest had been destroyed; the trees were lying on the ground all pointing away from a central position; 80 million trees over a 2,000 square kilometre region were flattened or burnt. Explanations for the Tunguska event have ranged from mini-black holes to alien space craft, yet all of the evidence is consistent with a comet or asteroid the size of an apartment block that exploded in our atmosphere several kilometres above the surface of the forest.

Asteroids larger than 1 kilometre strike the Earth's surface about every million years. Objects this size slamming into the Earth's surface can affect climate globally and lead to enormous destruction. Statistically, we are as likely to die from a giant impact as we are from being in an aeroplane crash. Flying in an aeroplane is pretty scary to many people, including me, and for no statistically valid reason. It is actually reasonably safe. It is probably the feeling of not being in control and the fear of knowing you are most likely to die during the several minutes before impact. An asteroid larger than a couple of kilometres across hits the Earth about once every five million years. Such an event could lead to the death of a large fraction of the human race. Therefore, you have a one in five million chance of dying from an asteroid impact in any given year. Over your lifetime, your odds of dying this way are about one in 66,000. Your odds of dying in a single 1-hour aeroplane flight are about one in a million, so if you fly several times per year you have the same likelihood of dying as from the global effects of a giant asteroid strike. The chances are that such a large asteroid will not collide with Earth during our lifetimes. But it will happen to our descendents if our species survives for long enough.

The last major incident that led to mass extinctions across the planet occurred some 65 million years ago and is commonly known as the K-T extinction event. It is associated with a geological signature known as the K-T boundary, a thin band of sedimentary rock that is found all across the Earth. K is the traditional abbreviation for the Cretaceous period derived from the German name *Kreidezeit*, and T is the abbreviation for the Tertiary period. It was known that the rock had been deposited on the Earth's surface at the same time as the last mass extinction, but no one had been able to determine how the rock came to be there. In 1980 a research team including the physicist Luis Alvarez, discovered that the K-T boundary rock was found to contain one hundred times the usual abundance of the element iridium, which is very rare on Earth but is commonly found within asteroids. To cover the planet with debris, they calculated that a 10-kilometre-sized asteroid must have hit the Earth 65 million years ago. The search began for an ancient impact crater that was likely to be several hundred kilometres across.

Evidence from the fields of palaeontology, geochemistry, climate modelling, geophysics and sedimentology shows that the Chicxulub crater lying off the Yucatán peninsula was probably the impact event that caused the K-T boundary layer and ended the rule of dinosaurs. The impact site is largely buried under the ocean and land, but detailed maps of the area reveal the roughly circular 200-kilometer crater, which is littered with shocked quartz and tektites. Similar quartz crystals were first found at nuclear bomb testing sites; their unique crystalline structure is created as rock is melted under extreme pressure and temperatures. It is commonly found in impact craters but also across the entire planet over the K-T boundary.

The Tsar Bomba took one kilogram of matter and turned it into pure energy. The mushroom-shaped cloud from the explosion rose to seven times the height of Mt Everest, over 60 kilometres above sea level. It released an amount of energy similar to the volcanic eruption in Krakatau, which destroyed several Indonesian islands and created sonic blasts that were detected travelling around the entire planet for several days afterwards. Both of these events pale into comparison with the energy and devastation created by the asteroid that created the Chicxulub crater. The asteroid was about 10 kilometres across, the size of a city, and would have impacted the Earth at a speed of about 100,000 kilometres per hour. The energy released by stopping something this large moving at this speed is huge—the equivalent of exploding one million Tsar Bombas!

Our atmosphere does little to protect us from the giant asteroids that have wandered through the solar system for billions of years. The kilometre-sized objects pass through unscathed in a matter of seconds. If they were to land in the ocean, they would immediately vaporise millions of tonnes of water, leading to global cloud covering that would obscure the Sun. A kilometre-high tidal wave would devastate all the surrounding continents. The ocean does not protect the Earth beneath and the asteroid would barely feel the water on its way to the ocean floor, the crust of our planet. The resulting shock waves would trigger global earthquakes and volcanic eruptions. Debris material from the impact crater would be ejected high into the atmosphere but it would eventually fall back to Earth in a rain of deadly shooting stars that would land all across the planet, igniting wildfires. The dust in the atmosphere would remain for years, obscuring the Sun and interrupting the photosynthesis of plants, causing most of them to die. The entire surface of the planet would be covered by a thick layer of dust and ash, creating a harsh environment for plants and animals for many years.

All of this, combined with the volume of carbon dioxide (CO_2) released during the destruction of carbonate rocks, would result in a runaway greenhouse effect.

Global climate change would ensue. The loss of plants and plankton that depend on photosynthesis would lead to the extinction of many herbivorous animals, and consequently predatory animals would perish. Omnivores would have the best survival rates, feeding on detritus, non-living or decomposing organic material filled with micro-organisms. Creatures in streams and on the ocean floors which feed primarily on organic waste would be the least affected. The largest air-breathing survivors of the Chicxulub event were related to the crocodile family. They can survive months without food and their young are small and grow slowly, feeding on dead organisms for the first few years.

Giant asteroids are not likely to destroy all life. Rain would fall and the sky would slowly clear. Seeds can lay dormant for hundreds of years awaiting the right conditions to germinate. Life would be set back, but it would recover, with new creatures evolving to take over the niches left behind by extinguished species. Perhaps such events accelerate the diversity in life forms; after all, dinosaurs did not invent the wheel, and they may have prevented the evolution of more intelligent species that might have done so. Primates, the biological order which includes apes and monkeys, did not appear until shortly after the Chicxulub event.

The threat of giant impacts was highlighted when comet Shoemaker–Levy was discovered in 1993 to be on a collision course with Jupiter. The comet had already been broken into at least 20 pieces by the gravitational field of our largest planet, but some of the pieces were still several kilometres across. The impact took place in 1994 and was observed by all the major astronomical telescopes and satellites. It hit Jupiter travelling at 60 kilometres per second, creating a "fireball" with a temperature of 24,000 degrees centigrade that reached 3,000 kilometres in height.

Astronomers can now detect about 80 percent of all Earth-bound asteroids larger than a kilometre which could lead to a global catastrophe. Within a decade or two we should have the sky-scanning capability to detect all asteroids larger than about 100 metres. Given enough advance warning of a large object on a collision course with Earth, we also have the technology to deflect them from our path. This early detection and damage control would be a remarkable accomplishment for our species.

There are many ways in which we could divert an asteroid from its collision course with Earth or even destroy it completely. Sending our arsenal of nuclear weapons to blast it to pieces sounds like a good idea, and in principle this is the best that we could realistically achieve in the near future at short notice. Scientists have devised a weapon that could wipe out a significant fraction of life on Earth, but the same technology could be essential in saving the planet from a future giant impact. However, many asteroids and comets are loosely held together "rubble piles". An explosion would only break the object into numerous large fragments that would rain down on Earth, causing similar devastation. Comet Shoemaker–Levy is such an example. We would have to ensure that the incoming object was vaporised into enough small fragments that our atmosphere would then protect us from this lesser debris.

Alternatively, if the nuclear explosions could take place off the surface of the asteroid, we could perhaps deflect its trajectory just enough so that it would miss our planet. As Earth orbits the Sun, it moves a distance equal to its own size in just seven minutes. If an asteroid is observed and predicted to collide with us, we need to change its arrival time by at least this amount for it to fly safely past. This is effectively the same principle behind the method of nuclear pulse propulsion for sending a spacecraft to the stars described earlier. Orbital deflection of the asteroid could also be achieved by hitting it with a large-enough object, such as a massive spacecraft. If there were enough advance warning, we could send a smaller spacecraft to land on its surface and install a conventional rocket propulsion system, gently steering its trajectory away from our planet. However this is accomplished, it will certainly be necessary at some point in our future to ensure the long-term survival of our species.

24 hours + 02:00

The Sun Shines Brighter, Water on Earth Boils

As the Sun turns mass into energy by nuclear fusion, a fairly obvious consequence is that its central core is becoming less massive. As a consequence, the pressure drops and the core of the star contracts and heats up as gravitational energy is converted to the kinetic energy of the atoms. The increase in temperature from the gravitational contraction is much larger than the temperature decrease resulting from the drop in pressure. This gives the counter-intuitive result that, as the star loses mass, its temperature increases. The consequence is a higher rate of nuclear reactions and the luminosity, or power output of the Sun, slowly and steadily increases over time. Furthermore, the higher internal temperature will cause the outer regions of the Sun to expand. Thus, as time ticks on, the Sun is getting larger and more luminous, whereas in the past it was smaller and fainter.

When the fusion reactions at the Sun's core started 4.58 billion years ago, it shone with a luminosity that was about two thirds of today's value. Luminosity is simply a measure of the total number of photons leaving the Sun. The temperature on Earth is proportional to the number of photons that hit its surface. Therefore, 4.58 billion years ago the temperature on Earth would also have been lower by two thirds of today's value. That is the temperature measured on the Kelvin scale where zero Kelvin is absolute zero or −273.15 degrees centigrade, the coldest anything can ever be. At this temperature all motion literally stops. The consequence of the Sun having been cooler is that, when the Earth formed, it would have been 90 degrees Kelvin colder than it is today, far below the freezing point of water. Such conditions would not have been good for the development of life, which requires stable warm temperatures. So why do we have evidence for liquid water on the Earth's surface four billion years ago when its surface should have been frozen? In fact, the geological evidence indicates that average surface temperature on Earth has been reasonably constant over the past few billion years despite the ever increasing amount of sunlight it receives.

It is possible to accurately measure the temperature on Earth over the past million years by looking at the structure and composition of long cores of ice drilled out of ancient glaciers. The ice from the bottom of the glaciers formed from frozen rain and snow that fell millions of years ago, and it contains a record of the conditions at that time. The ice is layered, like a set of tree rings indicating a yearly weather cycle. Air trapped inside the ice for a million years in tiny bubbles can be analysed for various isotopes which can then be compared, for example the ratio of oxygen-18 to oxygen-16. The neutron-rich and heavier oxygen-18 evaporates more efficiently from the surface of the ocean when it is warmer. By measuring its abundance, we can reconstruct the historical record of the climate of the Earth. Over longer timescales we can look at the geology of strata of sedimentary rocks to infer the global mean temperatures.

The reason for this apparent paradox is that the temperature on Earth has been regulated by the greenhouse effect, which has maintained hospitable conditions on our planet over most of its history. The mean temperature on the surface of the Moon is about −19 degrees centigrade, well below the freezing point of water; this is the temperature that any rocky body should have at a distance like that of the Earth from the Sun. The mean surface temperature on Earth is actually +15 degrees centigrade, thanks to the greenhouse effect provided by our insulating atmosphere. Most of the photons from the Sun travel straight through our atmosphere and reach the Earth's surface, where they are absorbed by the oceans or the land. This warms the surface of our planet, which radiates the energy back, but in the infrared wavelengths, to which our eyes are not sensitive. This is the warmth that you feel when you put your hand above a warm rock that has been lying in the sun. Whereas the atmosphere is transparent to photons in the optical wavelengths, it is opaque to infrared photons, which scatter off the lower atmospheric layers and are reflected back onto the surface of the Earth. The heat is effectively trapped around the planet rather like an oven. On the Moon, which has no atmosphere, infrared photons escape back into space, which is why the surface temperature of the Moon is much lower than that of the Earth. The efficiency at which the atmosphere can reflect and trap infrared photons is sensitive to how many molecules of water, CO_2, methane and ozone are present.

For at least the past four billion years, the greenhouse effect has been beneficial for the evolution of life. In fact, the Earth's temperature has been remarkably stable for 2,000 years, not varying by more than a third of one degree over any 50-year timescale. However, over the last 50 years the temperature of our planet has increased by almost one degree. The evidence is very strong for the case that the recent rapid increase in the Earth's mean surface temperature is a direct consequence of the increase in CO_2 in our atmosphere—the consequence of industrial activity and man-made pollutants. We have yet to appreciate just how sensitive our global ecosystem is. Our species needs to be very careful not to trigger a global catastrophic climate change in the next microsecond after midnight.

The Earth moves on a slightly elliptical orbit, varying in distance from the Sun by five million kilometres over the course of a year, which seems like a lot but is only a 3 percent difference in its distance. This variation has little effect on the Earth's yearly variations in climate from summer to winter, which are due to the tilt of the Earth relative to the Sun. In fact the Earth is farthest from the Sun during summer in the Northern Hemisphere.

If the mean distance of the Earth to the Sun were to decrease by just 10 percent, the number of photons landing on the surface of the Earth would be 30 percent higher, leading to a mean temperature in excess of 100 degrees centigrade, the boiling point of water. The atmosphere and the oceans would evaporate as steam. If the Earth moved on an orbit that was over 40 percent more distant from the Sun, water on its surface could only exist in the form of solid ice. The optimal shielding from the greenhouse effect is included in this estimation. This region is the habitable zone of our solar system. Even so, it does not mean that life could not have developed in places outside of this region, for example within Saturn's icy water-filled moon Enceladus.

Although our Earth orbits in the habitable zone today, the temperature at the centre of the Sun is slowly increasing, and its luminosity, too, is growing at a rate of about 10 percent every billion years. In just one billion years from now the temperature on the Earth's surface will be 30 percent higher, which is sufficient to evaporate our atmosphere and boil the water from our oceans. According to our cosmic 24-hour clock, the Earth will no longer lie within the habitable zone and all life on its surface could die already by 2 am tomorrow.

A New Orbit for the Earth

If we wish to remain on the planet for a further seven billion years, we have to literally move the Earth into a new orbit that slowly increases its distance from our evolving Sun. This could be accomplished, maintaining a stable climate on Earth such that in a billion years from now it would be 50 percent further away from the Sun. I already mentioned that the orbital energy of the Earth is enough to supply our current power requirements for longer than the age of the universe. To move the Earth to 1.5 times its present distance from the Sun would require about one third of that amount. That sounds a very difficult problem, and in fact it is hard, but perhaps not impossible! The scientists Don Korycansky, Greg Laughlin and Fred Adams have already developed a plan. The principle is based on Newton's third law of action and reaction and employs the same technique used to accelerate spacecraft to high speeds in our solar system, so-called gravitational assists. The idea is to extract energy from the orbital motion of the largest planets—a vast reservoir of energy.

Voyager 1, launched in 1977, still holds the record for the fastest-moving spacecraft. It is leaving our solar system at about 60,000 kilometres per hour, but it did not start off with that speed. It was manoeuvred into a trajectory that would make it fly past Jupiter at a distance of just twice the diameter of Jupiter away and behind the planet relative to its orbit around the Sun. If Jupiter were not moving, the spacecraft would approach and leave Jupiter at the same speed. However, as Jupiter is moving around the Sun, its gravitational force literally drags the spacecraft along as it passes behind it, accelerating it and doubling its speed. Of course, you do not get energy for nothing. Because Jupiter gave some of its energy to Voyager 1, Jupiter slowed down as a consequence. However, since it is so massive, its change in speed was negligibly small—in a trillion years Jupiter will be about 30 centimetres behind where it should have been had Voyager not flown past.

We could move our entire planet into a new orbit using the same technique. The idea is to manoeuvre a comet or asteroid into an Earth-crossing orbit—one that passes in front of the Earth's motion around the Sun at a distance as close as 1.5 Earth radii away. It would need to pass in front of the Earth, and its gravitational force would lead to a slight increase in the speed of the Earth, causing the planet to move into a slightly larger orbit at the expense of the comet slowing down. A single passage of a 100-kilometre object would increase our distance from the Sun by less than 100 kilometres—a tiny amount given the 150 million kilometres that separates us from our star. We have to move the Earth at least 70 million kilometres farther away than it is already, which will take about a million encounters. These need to be precisely calculated and engineered through careful monitoring and steering of the asteroid. Since it would lose energy during each encounter with the Earth, it would be placed and kept on an orbit that carried it between the Earth and Jupiter, passing behind Jupiter's orbit to extract some of its energy. The asteroid is a device to steal energy from Jupiter's motion around the Sun and pass it on to the Earth: Jupiter would lose the energy that the Earth gained. But Jupiter has a lot more energy to begin with because the more mass you have, the more energy you have, and Jupiter is a thousand times as massive as our Earth. After a billion years, during which the Earth will have moved 50 percent farther than its current orbit, Jupiter would end up only 1 percent closer to the Sun.

This would take a good fraction of the remaining billion years to accomplish, but we have the time. And although it sounds outrageously difficult, it is all perfectly possible. We do not need to start now; we could wait a million years since it will be much easier to accomplish with a little advancement in our technological capabilities. We also need to reach a highly stable and long-lived society, devoid of wars and terrorism. Otherwise, we could be faced with the threat of some crazed individual adjusting the orbit of the asteroid just to put it on a direct collision course with the Earth!

24 hours + 07:00

Cosmic Collisions

If somehow the Earth manages to avoid extinction from collisions with asteroids, we should also consider whether it might collide with other stars or even other galaxies. This may sound an unlikely problem, but by 7 am in the morning (four billion years from now) the Andromeda galaxy will collide with our Milky Way.

The Andromeda galaxy is similar to ours in appearance and size. It is the only galaxy in the Northern Hemisphere which is visible to the naked eye. It is best to look for it in autumn at dusk—when the Moon is still below the horizon. You can find it by looking for the constellation Cassiopeia. It is like a squished W; the right-hand V part of the constellation points towards Andromeda, which is about three times the height of the V away. It is very faint, but with patience you will see a fuzzy-looking patch of light that extends to about the size of the Moon. With binoculars or a telescope you can resolve the fuzzy patch into a beautiful spiral galaxy that is about two million light years away. If you can see Andromeda, you are looking at the light which was emitted by its stars over two million years ago, at around the time our ancestors started to walk upright on two feet.

You might be puzzled as to why Andromeda is heading our way, when I already stated that space is expanding and that the distances between galaxies are growing. A galaxy is like a piece of the universe that contained more mass than average—enough mass such that gravity has reversed the cosmic expansion and matter from a large region has come together to form a galaxy. Similarly, if two galaxies are close enough together, they can also reverse their expansion and start to head towards each other. That is just what is happening to Andromeda and our Milky Way; several billion years ago Andromeda stopped moving away from us and started to move towards us. The very first project I undertook as an astrophysicist was to calculate the future trajectories of the Milky Way and Andromeda.

Travelling to Andromeda at the speed of light would take two million years, not a journey that is likely to be undertaken. But if we wait we will not need to—Andromeda is coming to us! It is travelling towards us at 430,000 kilometres per hour. Moreover, it is speeding up as it approaches, pulled quicker and quicker by the gravitational attraction of our Galaxy. It will collide with our Milky Way in about four billion years from now at a speed of about two million kilometres per hour. The collision and resulting coalescence of the two galaxies will lead to the formation of a single new "elliptical" galaxy—shaped more like a boiled egg than the fried egg it resembles today—that will be twice as massive, spherical and effectively dead. The stars will no longer rotate in a well-ordered disc; because of the violence of the collision, they will be thrown in all directions—some right out of the galaxy. There will be no orderly Milky Way. After the collision, the end state will be a more or less spherical distribution of stars moving randomly in all directions and uniformly distributed across the night sky. Although most of the stars will travel straight past each other, that will not be the case for the interstellar gaseous medium. The two enormous layers of gas will collide, generating high temperatures followed by rapid cooling and a violent cascade of turbulence that will trigger vast numbers of new stars to form.

When we look out at the universe through a telescope, we can find similar collisions between distant galaxies taking place. We observe galaxies at different stages of collisions: Some are just approaching each other, whereas others may be seen in the middle of a cosmic dance. The images capture their current state, which appears frozen in time. Beyond the closest stars we rarely observe things changing over the lifetime of a human, or even over the lifetime of our species. Cosmic timescales are simply too long. That some galaxies collide is a natural consequence of gravity pulling nearby things together. If there were enough matter in the universe, everything would converge. But our inventory of cosmic matter tells us that there is not enough mass in the universe for this to happen. Andromeda will be the last big galaxy to collide with our Milky Way.

A pioneering attempt to "simulate" the motions of colliding galaxies was made by Erik Holmberg in 1941 using 80 light bulbs arranged into two "galaxies". This was before digital computers, but Holmberg realised that the light intensity falls off just like gravity. Consequently, to calculate the effective "gravitational force" on one of his stars due to all the other stars, he could just place a photocell in the appropriate place. On calculating the light intensity he could work out which way the light bulb would move. He would then go to the next light bulb and repeat the measurement. Today we can construct virtual galaxies in our supercomputers consisting of billions of points that represent the stars. Just like Holmberg's mechanical galaxy simulation, the computer calculation proceeds in small intervals of time, over which we calculate the position to which each star has moved as it is influenced by the gravitational force of all the other stars.

We can simulate in exquisite detail the motions of two galaxies over billions of years. One of the movies on my website, www.astroparticle.net, shows the future of our Milky Way as it collides and coalesces with Andromeda. The collision appears to be particularly violent, but it is actually not so bad. Out of several hundred billion stars, it is likely that only a few will directly collide. Stars are tiny compared with the space between them in a galaxy. There is a chance that a rapidly passing star from Andromeda might perturb the orbits of the planets, sending them closer to the Sun.

24 hours + 13:13

The End of Our Earth

Just after lunchtime tomorrow, according to our special clock, the Sun will finally die. By this time we will have to organise a mass migration of our species from the planet. A true Noah's Ark would need to be constructed to take what is left of life on Earth to another planet that is orbiting a younger star. In five billion years the evolution of our Sun will start to speed up since all of the hydrogen in the central core will have been used up in the fusion reactions. At this point, the nuclear burning of hydrogen continues in a shell which moves outwards through the star, further increasing its temperature.

Accurate stellar modelling of the Sun shows that in 7.6 billion years it will reach its maximum luminosity, which is over two thousand times higher than it is today. The temperature at the core will have risen from 15 million Kelvin to a massive 100 million Kelvin. At this temperature the helium nuclei will have enough energy to fuse together to form carbon and oxygen nuclei. These new fusion reactions will continue to turn mass into energy, which will cause the outer parts of the Sun to rapidly expand as it enters a brief "red giant" phase. In just a few million more years the Sun will expand to 250 times its present size, engulfing the Earth, which today orbits at a distance of 150 million kilometres from the Sun, only 215 times the Sun's radius away. The outer temperature in the envelope of the Sun at this stage will still be over 3,000 degrees centigrade, which would be sufficient to melt the Earth. It would turn back into a giant sphere of molten rock, just as it was shortly after it formed.

The outer layers of the Sun will blow away into space, and the Sun will enter a brief period as a "planetary nebula", a giant cloud of expanding gas illuminated by the remaining core of the star. Planetary nebulae actually have nothing to do with planets. The name comes from early astronomical observations of the night sky which revealed these colourful objects that resembled planets. This late phase of stellar evolution lasts about 10,000 years, after which time the gas disperses into the galaxy. Planetary nebulae are among some of the most bizarre and beautiful images that you may have seen taken by our telescopes.

If we have not managed to move the Earth outwards, in 7.6 billion years it might be completely incinerated within the outer layers of the Sun. Such an event would not destroy the Earth as completely as the Death Star from Star Wars, but it would come close. However, we must also take into account that the Sun is losing mass as it ages, not only by turning mass into energy but mainly through radiation pressure that constantly blows away the outermost solar material into space. At this time our Sun will be 30 percent less massive than it is today. As the Sun loses mass, the gravitational force on the Earth decreases; this results in the Earth slowly moving farther away from the star into an orbit just beyond the Sun's scorching outer edge. If the Earth is not charred molten rock by then, we would have a spectacular grandstand view of its death following its 12-billion-year life. However, despite the Sun providing the energy for life for our planet, it does seem intent on completely destroying the Earth. Eventually, the outer reaches of the Sun will drag our planet into its fiery midst. The molten Earth will sink rapidly to the very centre of the dying star, ultimately crashing onto the surface of its dense core and being torn into its component atoms by the intense gravitational forces at the Sun's centre. The Sun would barely react as it swallows the Earth-turned-specks-of-dust.

24 hours + 13:14

The Death of the Sun

Once all the helium at the centre of the Sun has been converted to carbon and oxygen, the temperature will not be high enough to create any heavier atoms and the energy source will switch off. There is nothing to prevent the gravitational forces from pulling the matter into a denser and denser state. If there were no force to oppose the gravity, it would not stop until all of the atoms were pulled so close together that they form a black hole. However, gravity must overcome several barriers to bring matter to a state so dense that light itself cannot escape.

When the Sun stops its energy production by nuclear fusion, the gravitational collapse will resume at full strength with nothing to oppose it. The central material will collapse into its final structure, namely a "white dwarf". The atoms become so closely packed that the electrons provide an opposing pressure called quantum degeneracy, which is a fundamental principle of particle physics that tells us that two electrons with the same energies cannot occupy the same space. The white dwarf core begins its life hot and glows brightly with the residual energy of the star it once was. But as it slowly radiates away its heat energy, it becomes dimmer and dimmer.

We are familiar with the normal atomic state of matter that exists on Earth. But we have little experience with the extreme states of atomic matter that can and do exist in our universe. A cubic metre of air weighs about one kilogram. A cubic metre of water weighs 1,000 kilograms (one tonne). The densest naturally occurring substance on Earth is the element osmium—it is so durable and hard that it is used in the writing tips of ballpoint pens. A cubic metre of osmium would weigh over 22 tonnes. These different densities arise just because of the spacing between the atoms. In a metal the atoms are arranged in a tightly packed crystalline configuration. Normal atomic matter can also be compressed into a denser configuration such that the spacing between the atoms becomes smaller and smaller. For example, we normally think of hydrogen as a diffuse gas; however, a cubic metre of hydrogen from the core of the Sun would weigh 150 tonnes. This is because of the immense gravitational pressure from the outer layers of our star that compresses the gas at its centre to such a high density.

When the Sun runs out of fuel for nuclear fusion 7.6 billion years from now, there will be no energy source to maintain the random motions of particles—and that's what counters the inwards pull of gravity thus maintaining the Sun's size and shape. Gravity finally takes over and most of the mass of the Sun will collapse and will keep collapsing until the electron-degeneracy pressure prevents the collapse proceeding any further. By this time our star will be squeezed into a volume the size of the Earth—the nuclei of the carbon and oxygen atoms live within a sea of electrons that provides the pressure to prevent further collapse. A thimbleful of a white dwarf would weigh more than a tonne, and a cubic metre of this strange electron-degenerate state of matter would weigh a million tonnes! When the first observations of the masses and densities of a white dwarf star were made in 1914, the numbers were simply not understood. A decade later, application of the new theory of quantum mechanics showed exactly how matter should behave under these conditions.

Of the hundred stars closest to the Earth, there are eight white dwarfs. As these objects cool down slowly over time, the most carbon rich will eventually cool and crystallise to form a structure that closely resembles a diamond the size of the Earth! Do we need a better reason to explore our Galaxy than to witness its incredible contents?

Atoms Are Mainly Empty Space

How is this possible? Well, atoms are mainly empty space, a remarkable fact given that a normal object like a table looks and feels solid. Surely if it were made of tiny atoms that have lots of space between them, we should be able to pass our hand right through the table? I will discuss this further in the final chapter when we look at our perception and reality of the universe. To give you a sense of just how empty an atom actually is, if it were possible to take a normal hydrogen atom and rescale its size by expanding it to the size of the Earth, the proton and neutron that make up its nucleus would together be the size of a football stadium and the electron would be just 10 centimetres across and located on the opposite side of the Earth, 12,000 kilometres away. The rest of the hydrogen atom is just empty space.

A white dwarf is not the ultimate dense state of matter, far from it. In 1931, the Indian astrophysicist Subrahmanyan Chandrasekhar calculated that the maximum mass of a white dwarf should be no more than 1.4 times the mass of the Sun. This is precisely what is observed when we measure their masses today. If the oxygen- and carbon-rich core of a star were more massive than this, its gravity would overcome the electron-degeneracy pressure. In this case the atoms can be compacted even closer together, so close that the nuclei are effectively touching. The end state is a neutron star which has the density of atomic nuclei. It is the strong force that keeps the proton and neutron separate in the nucleus, but it is the neutron degeneracy pressure that prevents the star from collapsing further.

The entire core of the star with over 1.4 times the mass of our Sun is compressed into a volume the size of Mont Blanc! A cubic metre of a neutron star would weigh an incredible thousand trillion tonnes—a thimbleful weighs a billion tonnes. If an object were dropped from a height of one metre above the Earth's surface, it would take about half a second to hit the ground, at which time it would be travelling at two metres per second. If you could do the same experiment dropping an object from one metre above the surface of a neutron star, it would take just one microsecond to hit the surface and it would be travelling at two million metres per second. The force of gravity from a neutron star immediately destroys any material falling onto its surface, emitting a burst of energy as it disintegrates and its motion is abruptly stopped.

Neutron stars rotate very fast: up to several hundred times a second. This is a consequence of the conservation of rotational energy that I mentioned earlier. From looking at how the sunspots move, Galileo realised that the Sun rotates about once per month. A star rotating at this speed and then collapsing down to 10 kilometres would spin a billion times faster and neutron stars have actually been observed spinning this fast. The rate at which they spin can be measured by studying pulsars, neutron stars which emit regular pulses of radiation in our direction. In the same way that the neutron stars spin faster and faster as they collapse, their magnetic fields also become stronger and stronger. The charged particles on the surface of a neutron star interact with its magnetic field. As a result, it gives rise to electromagnetic radiation—a beam of light from its north and south poles is emitted into our galaxy. Some of the pulsars wobble as they spin and we only observe the light when it is pointing towards us, rather like the beacon of a lighthouse. The milliseconds between the pulses of light are measured to be as regular as our atomic clocks. The universe gives us many ways to keep an accurate track of time!

Supernova

Even a neutron star is not the ultimate state of matter—that would be a black hole. Almost a thousand years ago on 4 July 1054 AD, a new star appeared in the sky which was as bright as the full Moon. It could be seen during the day for over a month and was visible at night for two years. We know this from the records of Chinese, Japanese and Arabian astronomers. Even the Native American Anasazi Indians carved a petroglyph in Chaco Canyon, New Mexico, which showed the crescent Moon alongside a very bright star in the same position as recorded by the Chinese. It is a mystery why no mention of this "guest star" was made by Europeans. Perhaps it symbolises the long period of scientific stagnation during which the dominant church imposed the view that everything in the sky was eternal and should be recorded as such!

Seven hundred years later the French astronomer Charles Messier began a painstaking program of observing the night sky looking for faint extended objects. He catalogued everything he observed that was neither a star nor a comet and listed 110 "nebula" objects, named M1 to M110, a notation that is still used today. The first object on his list was M1, a diffuse nebula in the same position as recorded by the Chinese astronomers. What Messier had observed was the expanding debris from the star, which exploded in 1054. Today the debris is over 10 light years across and it is still expanding from a central point at a speed of 1,500 kilometres per second. Just after the star exploded, the debris would have been travelling at speeds of 30,000 kilometres per second, which is 10 percent of the speed of light. We know this because we can look back at historical observations of M1 and compare the size of the nebula with today's images and estimate how quickly it has expanded into its presently observed size.

We are fortunate that it did not explode closer to the Earth, otherwise it would have been both spectacular and deadly. For a supernova to appear as bright as the Sun, it would need to be located just one light year away from the Earth. Within ten light years from us there are only a dozen stars, and none is expected to end its life soon. This is good because the blast wave of cosmic rays and high-energy photons across the energy spectrum would be rather detrimental to life on Earth. The star that exploded in the year 1054 was 6,000 light years from the Sun. There are millions of stars that are closer, and the star that is most likely to go supernova is Betelgeuse in the constellation of Orion. Betelgeuse is only about ten million years old, but it is 20 times as massive as the Sun and stars of this mass evolve very rapidly. It is about 500 light years from the Earth. One day it will light up the sky 100 times as bright as the supernova in 1054.

Supernovae are observed about once every hundred years in galaxies. Astronomers observed a star exploding in one of our neighbouring galaxies, the Large Magellanic Cloud, as recently as 1987. The last star to explode in our Milky Way was in 1604, known as Kepler's Nova. It was visible during the daytime for over three weeks even though it was at a distance of 20,000 light years from Earth. The one before Kepler's Nova occurred in 1572 and is named after the sixteenth-century Danish astronomer Tycho Brahe. Brahe was a wealthy nobleman, as were many of the first scientists. They had the time and money to think. He built an astronomical observatory from which he began to systematically study the night sky. From the age of 17 he began to carefully document what he saw. His observations and measurements of the bright new star in 1572 again challenged the prevalent idea of an "unchanging cosmos". Brahe tried to measure its distance using parallax, but it showed none, implying that it must be further away than the Moon and the planets.

At that time, the established model for the universe was one in which the stars and planets were embedded in concentric rotating spheres made of a transparent fifth element. These "celestial spheres" were thought to be moved by angels, but the outermost eighth sphere was immovable. It had been identified by the Christians as the home of god, of heaven. Tycho Brahe was as eccentric as many scientists today. He lost part of his nose in a duel with another student after arguing over who was the better mathematician. He hosted extravagant parties during which his guests would be entertained by his pet elk, which died one night after drinking too much beer and falling down the stairs!

How stars evolve and die depends sensitively on their initial mass. Any gas cloud that collapses into an object with a mass less than about 10 percent that of the Sun will become a brown dwarf, resembling a gas giant planet like Jupiter or Saturn. In these objects the gravitational collapse does not generate enough energy for nuclear fusion to start working. Brown dwarfs are failed stars and only glow dimly as they radiate away their heat energy obtained from their gravitational collapse and formation. Stars that have a mass between 1.4 and 10 times the mass of the Sun will burn through their fuel more quickly. Consequently, they live shorter lives, their cores collapsing into neutron stars.

About one in a hundred stars forms with an initial mass over ten times that of our Sun. In these stars the gravitational squeezing is so intense that their core temperatures exceed 600 million degrees. At this temperature the carbon and oxygen atoms begin to fuse together to create shells of heavier elements all the way up to iron. The internal structure of the star resembles an onion, with an iron core surrounded by a layer of silicon, then a shell of carbon and oxygen surrounded by the outer layers of helium and hydrogen. Stars cannot fuse together iron atoms to form heavier elements though, since that would require more energy than is released from the fusion process and there isn't enough energy available. At this point the hot core begins to cool. It cannot withstand the enormous gravitational forces and rapidly collapses—a violent implosion that once under way cannot be stopped. The quantum mechanical forces are finally overcome, and the massive iron core collapses at a speed of 100,000 kilometres per second, instantly creating temperatures of 100 billion Kelvin.

The gravitational energy released during the collapse is enormous and sufficient to create all the elements heavier than iron, including radioactive elements such as uranium and precious metals such as gold. A shock wave propagates outwards that destroys the star in a giant explosion—a supernova. And all of this happens in just a few seconds! The collapse of the massive iron core releases enough energy so that the supernova shines as brightly as an entire galaxy of stars. For a brief period the luminosity of the star increases more than a billionfold, releasing the same amount of energy in just a few days as our Sun will emit over its entire lifetime. There have been no supernovae anywhere in our galaxy since 1604, even though we should have had at least four. So we are statistically long overdue for some spectacular cosmic fireworks in our night sky.

The energy of the collapse is transferred outwards. But there is so much mass in the collapsing metallic core that gravity finally wins and the object becomes so dense that even light cannot escape from its surface. No other known force can stop the matter from collapsing to an infinitesimally small singularity. I do not particularly like singularities and infinities; but no matter the internal structure of a black hole, the final state is invisible from the outside.

Supermassive Black Holes

In the last decade astronomers have been monitoring the motions of stars that reside at the heart of our galaxy. One of the stars, called S2, is travelling at 5,000 kilometres per second around "something" that is at a distance of just 17 light hours away. Its entire elliptical orbit has been mapped, and whatever it is orbiting around is invisible. That's right—we see absolutely nothing there, and we have looked with all our telescopes and satellites in all parts of the electromagnetic spectrum.

The star S2 is moving so fast that it must be moving around something very massive, something that has enough mass to stop S2 travelling onwards through our galaxy. Just as we can use Newton's laws to calculate the mass of the Sun knowing the Earth's orbital period and the distance to the Sun, we can easily measure the mass of whatever it is that star S2 is moving around; it turns out to be something that has a mass that is four million times the mass of our Sun. That is four million solar masses within a region three times the size of our solar system!

There is no reasonable alternative to the theory that the star S2 is orbiting a very massive black hole. Once you have four million solar masses within this amount of space, it has to form a black hole—this means that there is so much gravitational mass within such a small volume that it causes spacetime to curve back on itself. Even light cannot escape its presence. The size of the black hole might be thought of as the radius within which matter and light would be eternally trapped. For our Sun to become a black hole, we would need to compress it into a region just three kilometres across. For our enormous galactic black hole, the event horizon past which light cannot escape is six light hours across; this is similar to the orbital distance of Pluto. As we observe the centres of other galaxies, we find evidence that most of them have similar or even more massive black holes at their centres. We do not know how these objects form, although we have some good ideas—this is an exciting area of research that we are working on.

Perhaps the most popular tourist destination for our future astronauts would be a fly-by of the supermassive black hole that lives at the centre of our galaxy. As we fly past the event horizon of the black hole, we would experience all the incredible effects of time, space and gravity at work. The event horizon is the point of no return—the ultimate thrill for a suicidal skydiver. To an outside observer it would appear to take you an infinite amount of time to fall into the black hole. However, you would perceive time normally until the difference in the gravitational force between your feet and your head would stretch you like strands of spaghetti, eventually ripping apart the molecules in your body.

Midnight Tomorrow

It is already the end of tomorrow. If we have somehow collectively come to our senses and avoided worldwide war, prevented the rapid rise in the Earth's temperature from global warming and avoided extinction by giant asteroids, life and humans can continue to live on our planet. If we can divert our attention and the waste of time and money from war to the advancement of knowledge, our species could have long since filled our galaxy. We would need to at least avoid the incineration of our planet by our own Sun in the early hours of the first day and then hope for the best when the rain of stars from the Andromeda galaxy passes through and merges together with our Milky Way. We would wake up the next day and find that our spinning disclike galaxy had been gravitationally restructured into a nearly spherical elliptical galaxy containing twice as many stars. After lunch our Sun would stop shining. By the end of the day very few new stars would be forming and most of the stars that are similar to our Sun would have since died. The cold dark remnants of white dwarfs, neutron stars and black holes would continue to orbit unseen through the galaxy. However, the smallest stars would be slowly burning their atomic fuel and could continue to shine for a while longer. Life could continue in our galaxy, but its energy resources would be dwindling. In the next chapter I will discuss the prospects for life to continue into the day after tomorrow and into the New Year, as all the stars within our galaxy die and fade while everything beyond our galaxy disappears due to a strange acceleration of the expansion of space.

Chapter 9
The Future of the Visible Universe

B. Moore, *Elephants in Space*, Astronomers' Universe,
DOI 10.1007/978-3-319-05672-2_9,

On those rare clear days in Seattle when it isn't raining, Mount Rainier provides an imposing presence in the distance. The massive volcano rises over four thousand metres above the surrounding landscape. It is also considered one of the most dangerous volcanoes in the world for its potential impact on human life if it were to become active again. The previous night in the College Inn pub we had come up with a new plan—we would leave after work and drive to the base at Paradise with the aim of climbing the mountain in the dark to reach the top by sunrise. This would ensure the most spectacular view at breakfast and maximum safety since the snow bridges over the crevasses would stay frozen solid in the cold night air. Our goal was to snowboard down at least part of the volcano, back to the car and make it to work the next morning. Descent by foot can be as tiring as the ascent, and we were looking forward to the ride three thousand vertical metres down the mountain. With crampons, ice axes and snowboards strapped to our backs, our spirits were high as we walked, our headlamps illuminating the glaciers that slowly drip down its sides. Mount Rainier has a topographic prominence higher than K2, on the border of Pakistan and China, and the long climb through the night was fun. It was about eight hours to the summit and one hour down, the most amazing hour of snowboarding ever. The journey home was the most dangerous part of the trip since I was so exhausted. At one point I was sure there was a truck headed directly towards me in the same lane. I slammed the brakes on and the car skidded to a stop. Luckily, Joachim was too tired to be scared anymore, so I told him about the near miss with the truck. He pointed out that the road ahead and behind was empty. . .there was no truck.

We made a similar plan a few months later, but this time it was the perfect cone-shaped Mount Baker. It was a cold night. Really, really cold. The temperature dropped well below –20 degrees centigrade in the early hours—but the exertions climbing up kept us warm. The previous day was warm and the entire mountain started a little slushy, but as the temperature dropped the snow froze solid—the entire mountain became a cone of solid blue, hard ice. Breakfast at 4 am watching the sun rise above the clouds is always special. In great anticipation of a magical journey back down to sea level, we strapped on our snowboards—it was a relief not to have to carry those heavy things anymore.

Thankfully for life on Earth, water has some unique properties that are different from most liquids. Most material in a liquid form takes up a larger volume than when it is solid. This is because at it cools and solidifies, the atoms are moving more slowly and they become closer together, more ordered and well packed. Water is special in that it expands as its temperature drops below four degrees centigrade. Frozen water is less dense than its liquid form, which is why icebergs float on the ocean. Water is made of two hydrogen atoms that connect to one oxygen atom, an asymmetrical structure that prefers to bond together eight molecules in a regular crystalline configuration that occupies more of the available volume. This is a good thing for life, which probably evolved in water. If water were to become denser as it froze (like most substances do when they turn from liquid to solid), the ice would fall to the bottom of oceans and lakes, eventually freezing them solid from the bottom upwards. Another strange property of ice is that it is remarkably slippery. Solid water is very strange in this regard. You do not slip while walking on solid wood or concrete, or on almost any other solid surface. It is commonly thought that ice is slippery because, when you stand on it, the pressure on the surface increases the temperature and melts a thin top layer of water, creating a frictionless smooth surface. This is not correct. A person standing on ice skates only generates enough pressure to raise the temperature of the ice by a fraction of a degree. The reason ice is slippery is not really known. But it may have an inherent fluid layer on its surface caused by the motion of surface molecules that have nothing above to bind to and so move around in search of stability. The layer of surface molecules is rapidly vibrating, creating a slippery liquid surface. When the ice is too cold, it stops being slippery. In 1912, Captain Scott's Antarctic exploration team perished on the return journey from the South Pole because the temperature became so low that the sledges would not move across the ice. It was like pulling them on concrete.

When we set off together down the mountain, we both immediately crashed and started to slip and slide down its frozen sides—the Sun's early morning rays had done nothing to soften the hard ice. Mt Baker had frozen solid like a giant cone-shaped ice cube. The steel edges of our boards were not even scratching the ice. We stopped, freezing at high altitude, huddled together waiting for the ice to soften. After a few hours, the temperature had barely risen by a degree, and we had to get back down the mountain—so we put our crampons back on and descended with our ice axes in hand and snowboards on the backpacks. It was a long and exhausting walk back down to the car.

We have journeyed through tomorrow, and our universe is now twice as old as it is today. The next events in the future of our universe take place on a longer timescale: over the course of the next year on our rescaled time. That is 365 times the current age of the universe or several trillion years into our distant future. Over the course of the next hundred days the universe beyond our own galaxy disappears out of sight. Not because the stars are fading in all the distant galaxies but because the space between the galaxies is expanding faster and faster. This accelerated expansion of space was discovered at the end of the twentieth century. And while its nature is unknown, it will have a dramatic effect on the future of our visible universe. By the end of the year, the universe will have faded out of sight anyway as the longest-lived stars stop shining.

The Attraction of Gravity

Our story begins with Isaac Newton, who showed that the inverse square law of gravity could explain the regular elliptical motions of the planets observed by Brahe and studied by Kepler. Newton also investigated other possibilities for how the force of gravity might behave according to how the distance between objects increases. He comments in *Principia*: "I have now explained the two principal cases of attractions: to wit, when the centripetal forces decrease as the square of the ratio of the distances, or increase in a simple ratio of the distances, causing the bodies in both cases to revolve in conic sections, and composing spherical bodies whose centripetal forces observe the same law of increase or decrease in the recess from the centre as the forces of the particles themselves do; which is very remarkable."

This is the only place in the three volumes of his *Principia* where Newton expressed surprise. What Newton is saying is that there is another possible force law that could give rise to the regular motions of the planets—a force that *increases* with distance between the objects—and that this force is exactly proportional to the distance between the planets and the more massive Sun. (This is known as Hooke's law.) These are the only two "force laws" that allow perfectly repeating elliptical orbits of the planets. And of these, only the inverse square law reproduces the orbits of the planets and their moons. Indeed, most other force laws would cause the planets to wander about our solar system moving in and out of the habitable zone. It would be even more problematic for life if the force of gravity fell off more rapidly than with the inverse cube of the distance. The planetary orbits would be unstable and would have long since crashed into the Sun or drifted off into deep space.

Newton was uncomfortable with the notion of "action at a distance" implied by his equations. In 1692 he wrote in a letter to a colleague: "That one body may act upon another at a distance through a vacuum, without the mediation of anything else, by and through which their action and force may be conveyed from one another, is to me so great an absurdity, that I believe no man, who has in philosophical matters a competent faculty of thinking, can ever fall into it."

Newton's universe was the sea of stars that we see in the night sky, and he realised that this was an unstable situation under an attractive law of gravity. After all, Newton knew that when apples depart from their branches, they fall to Earth. Indeed, he passed on this story to William Stukeley, who wrote a biography of Newton in 1752: "After dinner, the weather being warm, we went into the garden, and drank thea under the shade of some apple trees;. . .he told me, he was just in the same situation, as when formerly, the notion of gravitation came into his mind. Why should that apple always descend perpendicularly to the ground, thought he to himself; occasion'd by the fall of an apple, as he sat in contemplative mood." A static distribution of stars would be unbalanced and collapse together owing to their mutual gravitational attractions. He wrote in the *Principia*: "And lest the systems of the fixed stars should, by their gravity, fall on each other, he [god] hath placed those systems at immense distances from one another."

In 1895 the German astronomer Hugo von Seeliger also considered the stability of the apparent universe. He concluded that a modification to Newton's laws which became effective at large distances, namely a force that pushes, could give stability to a static distribution of stars, preventing them from collapsing together. Einstein was faced with the same problem when he wrote down the equations of general relativity and applied them to the dynamics of the entire universe. The resulting formidable but elegant "field equations" relate the curvature of space to matter and energy densities in the universe. However, Einstein's knowledge of the universe was no better than Newton's. In 1915 Einstein's universe was also a sea of stars and nebulae—the only additional knowledge he had was that the distances to the stars were known. It wasn't until 1927 that the expansion of our universe was revealed, when Lemaître and Hubble showed that the fuzzy nebulae observed through telescopes were actually distant galaxies speeding away from us.

Einstein's derivation of general relativity can be seen in his famous *Zurich Notebook*, which contains his handwritten notes from 1915.[1] The notebook begins from both the front and the back simultaneously; his calculations meet in the middle. The first page from the back has the sketch and designs for a complex children's puzzle. Then he delves into electrodynamics and relativity before writing down the "metric tensor", which describes a path in curved space! This is the beginning of further notes and equations in which he develops the theory of general

[1] The digitized copy is available here: http://www.alberteinstein.info

relativity—a unified theory of gravity that links together the geometry of space with the matter and energy that it contains.

The Repulsion of Space

Einstein, like Newton and von Seeliger, also considered the stability of the apparently static universe. He realised that it was mathematically correct for him to add a "constant energy" to his equations, one that increases proportional to the amount of space. This allowed for a stable and stationary universe but one that required an additional force that was a property of space itself. This constant energy term became known as Einstein's cosmological constant. Einstein is often quoted as saying that introducing the cosmological constant was his biggest blunder. That comes from a conversation overheard between Einstein and George Gamow. Perhaps what Einstein was really saying was that he should have predicted in advance of the observational data that the universe was expanding, rather than criticising his own idea of a cosmological constant. Either way, as we shall see shortly, Einstein was justified in including this term in his equations.

A few years later, Hubble would show that the universe was not a static stationary distribution of stars, but an enormous expanding space filled with galaxies. At that point Einstein's cosmological constant term was largely forgotten. However, it still remained a possibility that the vacuum of space itself contained energy and that energy could act in a very different way to gravity. The addition of the cosmological constant is rather like the linearly *increasing* force law that Newton had considered three hundred years prior to Einstein. In fact, if we take the classical Newtonian limit of Einstein's equations, we recover a force law that behaves like a combination of the two force laws that Newton originally showed could give rise to the planetary orbits.

The 50 years following Hubble were a rather confusing time for theoretical cosmologists as they attempted to determine the cosmic inventory of matter and energy. If there were sufficient matter in the universe, it could not only have caused the expansion to stop but could also have acted to drag everything back together, perhaps eventually recreating the conditions from which the universe first arose. Many people liked this idea since it gave rise to the possibility of a cyclical universe, a new expansion somehow re-emerging from the aftermath—an infinitely occurring universe. Scientific prejudice exists (and can last a long time), but it does not always prove correct. And prejudice alone will not withstand the test against observational evidence and scrutiny from other scientists. There was a large prejudice in favour of such a cyclic model, not philosophically, but because of the physics involved. Such a model can have a near perfect balance between the energy of the universe that drives its expansion and gravity which attempts to reverse its expansion, a balance that leads to a universe with a flat geometry.

We are used to thinking of space being the same in all directions. If you were to walk in a constant direction, you would simply get further and further away from where you started. But Einstein showed that space in the universe is distorted by its mass and energy content—it can be curved. As a two-dimensional analogy to a curved space, consider that an ant has no idea that it lives on a spherical planet. An ant thinks that its world is flat, thus the centre and exterior of its world are irrelevant. If the ant walked in a straight line, it would eventually travel right around its world and return to its starting point. Until then it would have had no idea that its space is curved. But it could have determined this using geometry since angles and distances within a curved space have different properties from those within a flat space. When space is flat, the internal angles of a triangle always add up to 180 degrees. If a clever ant drew a triangle with straight lines on the surface of its world, it would add up the angles and find a value larger than 180 degrees. The sum of those angles would be related to the curvature of its space. It turns out from the equations of Friedmann that the curvature of space is exactly flat if the energy density of the universe is just balanced at the critical value in which the gravitational force of the matter is exactly sufficient to exactly stop its expansion. This prejudice arises simply because scientists will usually seek the simplest theory or the most natural explanation for an observation. This thinking is often called Occam's razor or the law of succinctness. If that fails, the next step is to consider more complicated solutions.

In a classical and simplified sense we can ask what it means to balance gravitational and kinetic energy. If you stood on the Moon and fired a bullet upwards into space, it would fall back down to the surface. The speed of a rifle bullet is about one kilometre per second. To escape the Moon's gravitational field you would need to set the object moving at a speed greater than 2.38 kilometres per second, at which point the bullet would travel away from the Moon forever. At precisely 2.38 kilometres per second, the speed of the bullet would slowly approach zero in an infinite time. Its kinetic energy becomes completely converted into gravitational potential energy—a perfectly balanced system. Gravitational potential energy is the "potential" to do some work, to move something. It is negative by definition.

A flat space requires no energy to maintain and perhaps no energy to form. It is the ultimate free lunch and a nice idea because any model that describes why our universe appeared has the convenient starting point that it required no existing energy to do so. As astronomers continued to weigh the universe, they were unable to find enough matter that could balance the expansion. It looked like the universe had finite positive energy. Adding up everything we know, in visible matter plus the mysterious dark matter, there is always a shortfall by a factor of about 10 smaller than that required to balance the expansion. In the language of general relativity, this implied that space would be curved in a strange way and not flat—more formally, we would call it hyperbolic.

At the same time as the cosmologists were trying to measure the amount of mass in the universe, in the 1980s astronomers began to create the first three-dimensional maps of the distribution of galaxies in the universe around us. They were not randomly distributed in space but clustered together in giant sheet and filamentary structures. Some places had many galaxies, others had giant voids with very few galaxies. The initial conditions that could give rise to these patterns were not random. In fact, the distribution of galaxies in our universe today accurately reflects the initial conditions that are predicted to have been generated very early in our universe during the inflationary era, 10^{-30} seconds after the big bang. Those small irregularities in the amount of matter in a given place are correlated across large scales—like the complex network of ripples on the surface of a lake. That pattern is preserved in the distribution of galaxies around us today. Surveys of millions of galaxies across vast volumes of our universe show the same beautiful patterns, sometimes called the cosmic web. It resembles the neural network in the human brain.

Dark Energy

Structure in the universe is formed by the gravitational driven growth of those initial irregularities. The resulting pattern in the distribution of the galaxies around us depends mainly on the amount of matter in the universe and how fast it has expanded. In the early 1990s computer simulations revealed that in order to match the observational data, the universe must have a very low density of matter—a fraction of that needed to stop the universe expanding forever. Most of the theorists disliked this idea, and they were already re-invoking Einstein's cosmological constant as a form of energy that could balance the natural energy budget that we would like the universe to have—which leads to a flat geometry of space. It was not the first time that Einstein's cosmological constant had been reintroduced to explain some aspect of our universe, but it was the first time it began to be taken as a serious possibility.

In 1998 astronomers in America and Australia were independently attempting to measure the rate of expansion of the universe in the past by repeating what Hubble had done but using very distant objects that existed just a few billion years after the big bang. We can look back just because the photons of light emitted at that time have taken billions of years to reach us. Most scientists thought they would find that the universe was expanding faster in the past since over time gravity should slow down the expansion. They found the opposite: The expansion was slower in the past than today—the rate at which the universe is expanding is increasing! Some form of negative energy seems to be pushing space apart. There appeared to be exactly enough of it to make the total energy density of the universe a constant over cosmic time such that space is indeed flat. The 2011 Nobel Prize in Physics was awarded for this remarkable observational discovery.

To determine the expansion rate of the universe at different epochs, it is necessary to measure accurate distances to faraway objects. Fortunately, the universe has provided us with such a reference light source that always has the same brightness: the fierce glow of a star that ends its life as a supernova. Supernovae that all explode with the same brightness are not the isolated massive stars I described earlier, but a special type of supernova which occurs when a white dwarf core receives enough matter from a close partner binary star to tip it over the edge and cause it to collapse and explode. Because supernovae are so bright, we can use them to measure the distances to very remote galaxies that lie at the edge of our visible universe.

The rate of supernovae occurring in the Milky Way is about one every century. By observing hundreds of galaxies routinely, night after night, it is possible to find individual stars at random places and distances in our universe that end their lives in this way. We can also measure the red shifts of the supernovae, which tell us how fast they are moving away from us. Comparing the distances with the expansion velocities reveals a prominent deviation from a simple gravity-driven deceleration of the universe. The real distances between objects are larger than they should be according to Hubble's law. The data imply that, about five billion years ago, the rate at which the universe was expanding started to speed up. Such an increase in the expansion rate could be explained by Einstein's cosmological constant. This constant energy term could have been positive or negative, but instead of a positive term that could cause the universe to stop expanding, it turned out to be negative. The profound implication is that a given volume of space has some sort of internal energy that is always pushing and trying to make the volume larger. As the volume increases, this force increases. As a consequence, the universe begins to expand faster and faster. Since the physical nature of this component of our universe remains a mystery, it has been given the name "dark energy".

It was first thought that the energy driving the expansion of space might be linked to the virtual particles that pop in and out of existence in a vacuum. But that idea was quickly ruled out since the force they could provide was far too small. Theorists have come up with several alternative interpretations of these observations. One is that some kind of dynamic energy or "fluid" fills all of space, but opposite to normal matter and energy. That energy is called quintessence after the fifth element of the ancient Greek philosophers. Another possibility is that perhaps there is something wrong with Einstein's theory of gravity; a different theory of gravity may include a prescription for this strange property of space.

The effects of dark energy are far too weak to probe in the laboratory; its strength is negligible compared with the forces that we can measure on Earth. Even over a volume the size of the solar system, the effect of dark energy is tiny and thought to be immeasurable. It is very weak—like having 10^{-29} grams of matter per cubic centimetre that repels rather than attracts. It is nothing like normal matter, so we cannot just go and collect some to study. It is only across the vast distances between galaxies that there is enough "space" that we can begin to measure its properties and effects.

The latest observations of the expansion rate of our universe strongly support Einstein's vision of a cosmological constant and the idea that the amount of dark energy is simply proportional to the amount of space. We can view dark energy as the cost of having space; that is, it takes a certain amount of energy to acquire a fixed volume of real estate in space. When a large region of our universe doubles its volume, the energy present in the new volume (its negative pressure on space) is twice as large. As space expands, more energy is available to continue to push apart the galaxies as more space is created between them. The rate of creation of new space increases with time.

If you were to say that it sounds suspicious that two of the major components of our universe have the word *dark* in their name, I would agree. You might get the impression, and rightfully so, that we do not know what most of the universe is made of. For both of these components, we know how much exists, how it is distributed and what the basic properties are. We have good ideas as to their nature, and we are testing these ideas through laboratory experiments and our observatories on mountain tops and in space. The observational evidence for dark energy has been compounded by independent data from the cosmic microwave background from which we can accurately measure the total energy density of the universe. The relative contributions are now known to a precision of about 1 percent. Our universe contains 68.3 percent dark energy and 26.8 percent dark matter. Atomic material accounts for 4.9 percent (of which about 75 percent is hydrogen, 24 percent is helium and all the other atoms make up the small remainder). The energy density in photons and neutrinos is very small, approximately 0.01 percent, despite the fact that these are the most common particles in the universe. That is what our universe consists of.

Since it was discovered that the universe had a beginning, many people have wondered about its future. Does the universe also have an end? This is perhaps the first time in history that we can actually answer this question with a reasonable degree of confidence, not from speculation or from philosophical considerations, but from observations obtained in the last decade. If Einstein was indeed correct, and it certainly looks that way, then our future is clear. We know how the universe has behaved over the past 13.8 billion years and we can calculate its future. Dark energy did not play a very important role over the history of our universe, but it was always present and it gradually becomes stronger as the volume of space increases.

The impact on our future of Einstein's cosmological constant is rather bizarre. There is nothing to stop the universe from expanding faster and faster. Because the amount of space between us and other galaxies is continuously increasing, the wavelengths of light emitted by their stars are also stretched longer and longer. As more space is created, there is more dark energy and thus space stretches even faster—exponentially fast. The universe will be roughly double in size in another 13.8 billion years, which on our timescale is by the end of tomorrow. The light from stars in the distant galaxies will be stretched into the infrared wavelengths, invisible to our human eyes. As time ticks on, the expansion of space will continue and as a result the wavelengths of photons will be rapidly stretched to microwaves then radio waves. In about a trillion years (150 times the current age of the universe) all the galaxies will become invisible as their light becomes stretched into wavelengths so long that the photons are undetectable.

One hundred and fifty days from now in our time, that is based on our compressed "universal day," the entire universe beyond our own galaxy will be dark.[2]

[2] The effect of dark energy on our distant future was contemplated in a recent study by Lawrence Krauss and Robert Scherrer titled "The return of the static universe and the end of cosmology".

There will be no way to detect the presence of anything outside of it—the photons will have such low energies and long wavelengths that they cannot trigger a response in any physical detector. If there are any astronomers in the distant future, they will have no way of knowing that the universe was expanding and will not be able to tell that there was a big bang or a beginning to the universe. All they will see are the objects within our galaxy. Dark energy is not strong enough to expand the space within our Milky Way, which will still be held tightly together by gravity.

Future of Stars in Our Galaxy

During the first new day after the merger with Andromeda, the Milky Way will have reconfigured into a large elliptical galaxy and all the stars more massive than the Sun will already have completed their life cycle. There will be no more supernovae, and over the course of the year the galaxy will see its last star die as a planetary nebula. Our galaxy will be filled with nearly a billion neutron stars, 100 billion white dwarfs and an equal number of brown dwarfs. Brown dwarfs are failed stars that are not massive enough to generate the high temperatures required to ignite fusion. By this time they will all have cooled down and become frozen balls of ultracold hydrogen.

At about the same time as the distant galaxies become invisible, our galaxy will still glow with the light from an estimated 300 billion of the lowest-mass stars, which live the longest. These are red dwarf stars that formed with a mass just above the hydrogen-burning limit of brown dwarfs, which is about one tenth of the mass of the Sun. These stars mix their internal fuel through convection, in the same way as air circulates in a convection oven. They burn their hydrogen slowly, and their gravity is insufficient to compress the central regions to allow helium to form. The lowest-mass stars capable of igniting nuclear fusion will shine at a luminosity that is 1/10,000th of that of the Sun. Three quarters of all the stars in our galaxy are red dwarfs, and the smallest of these continue to shine for the entire year—about five trillion years. Like the Sun, as they evolve and burn their fuel, they become hotter and more luminous. Before they eventually complete their cycle of fusion and shut down, they spend a brief time shining as brilliantly as the Sun shines today and our galaxy glows brightly before it rapidly fades to darkness.

Gliese 581 is a red dwarf star lying 20 light years from Earth in the constellation of Libra. It has about one third of the mass of the Sun. In 2010 the Swiss astronomer Stéphane Udry found evidence for an extrasolar planet with about five times the mass of Earth orbiting Gliese 581 in its habitable zone. In fact, there may be at least six planets orbiting this faint star. In 2008 Ukraine's national space agency broadcast a high-powered radio signal towards Gliese 581 that contained 501 messages selected from a competition on a social networking website. Its strange contents will reach the star in 2029!

Despite their long lives, there are several reasons why it may be difficult for life to develop on extrasolar planets. Gliese 581 emits only one percent of the energy of our Sun. Therefore, a habitable planet would need to be ten times closer to it than our Earth is to the Sun to receive the same amount of energy. At this distance the planet would be "tidally locked" to its star, with one side constantly facing the star and one side in perpetual darkness. This tidal locking occurs as a consequence of the strong gravitational field which raises powerful tidal forces on its surface that dissipate energy, slowing down the spin of the planet until it spins exactly once during an entire orbit around the star. That is the same reason we only ever see one side of our Moon—it is tidally locked to the Earth.

As a consequence, one side of the planet orbiting Gliese 581 has constant daylight and is always warm; the other side is continually dark and cold. One side of the planet could be an enormous warm ocean of water where life could develop. If the dark side of the planet were frozen ice, it may be difficult for life to develop fire. But perhaps, in a narrow warm rocky band encircling the boundary where it is forever dusk, plants could evolve and an interesting ecosphere for life could advance—life forms that are perpetually awake. Recent models of these strange worlds have shown that they could develop a circulating atmosphere that could bring warm rain to the dark side. Each side of the planet might evolve very different ecosystems—but they would likely be interdependent in some way. Red dwarf stars radiate most of their energy in the red and infrared wavelengths since they are smaller and colder than our Sun. Creatures on their planets may have evolved eyes sensitive to these longer wavelengths—they would literally see warmth which is what we feel as radiant heat energy. Certainly, life may have developed very differently from life on Earth. What a fascinating prospect!

Whether or not they host life, red dwarf stars could potentially be used as power sources to enable the survival of advanced life for several trillion years. Shortly after the end of the year, 10^{13} years after the big bang, all the stars in our galaxy will have stopped shining. Most of the atomic matter will be inside the white dwarf remnants of stars which slowly cool down as their residual heat energy is radiated into space and the galaxy glows ever dimmer in longer and longer wavelengths.

Our universe will be plunged into yet another dark age. The difference between the dark ages imposed on science by religion during and after the Roman era and the astronomical dark ages of the early universe, is that this dark age will be eternal. Life may still continue to exist, but it will have to be capable of manufacturing energy from the remnants of stars that litter our galaxy, or the vast reservoir of energy stored within its central black hole. Could intelligent life evolve sufficiently to accomplish this? It would likely take a merger of mind and machine and the ability to use long-lived structures such as the diamond remains of cold neutron stars. Before getting to that in the final chapter, let us do a reality check of what we really understand.

Chapter 10
What We Do Not Know

B. Moore, *Elephants in Space*, Astronomers' Universe,
DOI 10.1007/978-3-319-05672-2_10,

It was a spectacular winter's day in the Glarus mountains, about an hour south of Zurich. Joachim and I were making an ascent of Silberen, an isolated peak in a beautiful location surrounded by tall mountains. We slid upwards on our mountaineering skis, which were adorned with special fabric skins that are smooth when you move forwards but provide friction in the opposite direction to prevent you from slipping back down the mountain. On the way down you simply peel off the fabric, lock in your ski boots and enjoy the spectacular descent. Joachim led the way, breaking track through the deep snow. There was no one else to be seen on the entire mountain. We reached the summit to witness the sun about to set behind jagged peaks that rose above the cloud-filled valley bottom. It was already getting late, but we expected the journey down to take half an hour compared with the five hours it took to slip and slide our way to the top. I followed, enjoying the entire ascent as the views opened up and became more spectacular the higher we got. At the top of the mountain you could see for many miles, a crystal-clear day. It was already late, the sun setting below the clouds that ringed the mountain peaks. The Moon and Venus shone brightly together. I was looking forward to my dinner date with Suzanne back in Zurich, so we locked down our ski boots and I asked Joachim how to ski.

"What?"

The question was rhetorical and short followed by

"You can't ski?!"

"Nope, I've never skied. I've only ever snowboarded down mountains with you. It can't be that hard. Just tell me how to do it and let's get down. It's getting late."

"Uh...like this," Joachim demonstrated, gracefully carving a couple of turns down from the summit.

It looked easy. I set off and was fine for a few seconds until I had to turn a corner which sent me face down into the snow. After a dozen ungraceful crashes, with skis and poles flying in all directions (I realised why I usually snowboarded), I gave up. Cursing and swearing I began the long walk back down, each step sinking in the soft snow up to my knees. But it was steep and gravity was on my side this time. It quickly got dark and very cold, so cold that the bottle of water in my backpack froze solid and my mobile phone battery refused to provide any power. Five hours later, walking in the light of the moon, we reached the car, the journey down through the deep snows as exhausting as the walk up. Suzanne, unable to contact us, assumed that we were lost and had given up on dinner plans. Skiing sucks.

Throughout this book I have tried to convey the incredible amount of knowledge that we have obtained about our universe and our physical world. The cumulative effort of scientists over several thousand years has led to this understanding. There is still a final chapter on this to come, but before contemplating the ultimate fate of the universe and of life itself, let us look from a more critical viewpoint at the knowledge that we have acquired and ask the question, What do we really know? I have made quite a big deal of the fact that, at this moment in our history, we have determined our place in space and time within our universe. And that is a truly incredible achievement from which I do not wish to distract in this chapter. But it is good to question things at an even deeper level and to reveal some of the basic things that we do not completely understand, which include space and time itself.

I would like to discuss our perception of the world around us and its reality. Our walls are mainly empty space and are transparent to certain particles. So what prevents us from walking or seeing through walls? Why do things obey fundamental "laws of physics" and where do these laws arise from? This takes us to questions that philosophers and scientists have debated for years and still attempt to answer today. It leads us through a discussion of the nature of particles and forces, to quantum mechanics and the arrow of time. Ultimately, all of this may shed light on whether our universe is deterministic and whether or not "free will" exists. These considerations may help us to answer another of those grand questions: What is the purpose of all this anyway?!

Space

Parmenides of Elea was the founder of the Eleatic school of philosophy in the fifth century BC. He had a strong influence on Plato and therefore influenced the whole history of Western philosophy. Parmenides thought that truth cannot be known through sensory perception and that only words and reason will result in an understanding of the truth of the world. Perception can indeed be deceptive. Zeno was one of his students and was perhaps the first to use proof by contradiction, *reductio ad absurdum* or reduction to the absurd, in which a concept or theory is disproven due to one of its predictions leading to an absurd consequence. Parmenides and Zeno were the founders of metaphysical logic, which attempts to answer questions such as, What is there? They questioned the reality of the world around us and believed that movement could not occur—any movement would require a void to move through, which would effectively be a nothing and a nothing cannot exist. As we shall see shortly, it does seem rather difficult to create or even imagine a true void.

The paradox of the nature of space was highlighted in Zeno's "dichotomy paradox". In order to travel a certain distance, say one kilometre, you must first travel half a kilometre and before that a quarter of a kilometre and before that an eighth and so on... By halving the previous distance there is always a smaller step to be reached and thus any journey could be considered as the sum of an infinite number of previously halved journeys...and thus any journey would require an infinite number of steps. Consequently, to reach any distance from the beginning would require an infinite number of movements. Zeno argued that this was absurd and that even the slightest motion was impossible and must be an illusion. Democritus and his mentor Leucippus thought hard about the structure of matter, noticing how materials could be mixed and how matter disintegrates into smaller pieces. To counter the Eleatic philosophers, they proposed that on the smallest scales there exist only "atoms" and "the void". Since motion was observed to occur, they argued that the void must exist.

Our understanding of the nature of "space" did not progress for another two thousand years until Descartes, Galileo and Newton again questioned how things moved and speculated about forces that attract all particles. It was the consideration of these basic questions that led to the understanding that we have today. Newton also based his theory of gravitation on the ideas of atoms and a void, that the planets moved through empty space around the Sun and that the force of gravity propagates through the space between all atoms in a way that depends on the amount of space between them.

You might think that Einstein would have submitted his doctoral thesis on the nature of space, time and gravity, since these are the subjects for which he is most famous. But no, it was on a technique called Brownian motion, which could be used to measure the sizes of molecules and atoms. The thesis is kept deep in the vaults at the library of the University of Zurich. It is a wonderful document to hold—the shortest thesis I have ever seen—just 16 pages of text! The sizes of atoms turned out to be much smaller than people had imagined. A sugar cube is about one centimetre on each side and contains about 10^{22} atoms; that is the same number of atoms as there are stars in our visible universe. Sugars are molecular chains of hydrogen, carbon and oxygen atoms. Each atom has an effective size of about 10^{-10} metres. The space between the atoms is about a thousand times larger than that, while the effective size of the protons or electrons in the atoms is at least a hundred thousand times smaller. Imagine expanding that sugar cube to the size of our visible universe, which means increasing its size by well over a billion billion billion times. Now, the mean distance between the atoms would be the same as if we spread out all the stars in the universe—about 10,000 light years. Each individual atom would have a size that is one light year across, and each particle in the atom would be the size of our solar system. Atoms do indeed seem to be mainly empty space.

So, just how small an object can we see with our microscopes and measuring devices? What does an individual atom or electron look like? Atoms are about a thousand times smaller than the wavelength of visible light, and we cannot "see" them by bouncing visible light off them. One could consider shining light of a smaller wavelength onto an atom, such as a gamma-ray photon that has a wavelength as short as 10^{-15} metres—that is about the classical size of an atomic nucleus. Unfortunately, a photon with such a short wavelength has a huge amount of energy. It might break the atom into its component particles, or at least knock the atom out of place.

The structure of an individual molecule was "imaged" for the first time by researchers at IBM's research laboratories in Zurich in 2009. They used an atomic force microscope that "feels" the electromagnetic forces very close to the molecules. The technique uses a tiny cantilever so small that its tip tapers down to a point that is about a nanometre across. The tip can detect the tiny forces that bind the atoms together. As it is moved over a molecule, the tip can construct an image of how the atoms are arranged. The scientists looked at a pentacene molecule made of 22 carbon atoms and 14 hydrogen atoms and confirmed that the atoms are configured in a perfect chain of hexagons.

We can explore some aspects of the internal structure of individual particles by crashing them together in particle accelerators like the LHC. Still, we have no idea what a single elementary particle, such as an electron, "looks like". It is defined mathematically, with properties such as mass and charge, and it gives rise to gravitational and electrical fields that permeate space and let other particles know it is there. In the language of quantum mechanics, electrons are little packets of energy waves that cannot be localised exactly—and at this point we have no familiar reference object.

If something is described only by equations and axioms, is it actually real, or could it be the manifestation of something else? If a single electron or proton is fired through a cloud of water vapour at high speed, it will ionise the gas and leave behind a visible trail. These entities seem to be "real" physical objects in this sense, and we can measure their properties with incredibly high precision.

We would like to understand our natural world in terms of things we have experience with, such as strings, balls or elastic bands. We can look at them or hold them, and our brains can understand them and interpret how they behave. Science began by attempting to explain things that we can see and observe in the world around us: the behaviour of water, fire, air and so on. We perceive our world through the reaction of our brains and senses to outside disturbances. Our senses are tuned for survival. As a result, we should not expect to be able to perceive any more than we need to. This perception that we have of our world is a narrow view. One could argue that it is a view that is not real and that solid objects do not really exist; atoms are mainly empty space after all. There is an awful lot more to our world than what we can perceive with our eyes. For example, the walls around us appear to us to be solid, but we can receive a radio signal inside a house. Radio waves and visible light are the same thing—they are both photons, only the photons have different wavelengths. It is the atomic structure of the walls that determines which photons can pass through, and those in the optical wavelengths are easily absorbed by concrete.

Our eyes are sensitive to the visible range of photon wavelengths which spans from about 370 to 740 nanometres. If for some reason we had evolved eyes sensitive to light with wavelengths in the range of 16–33 centimetres, our world would appear completely different. That is the range of wavelengths that our mobile phones use for communications, in the microwave part of the electromagnetic spectrum. Photons at those wavelengths can pass through our walls, enabling our mobile phones to work inside buildings. Our eyes would be blind to the buildings around us. We would see straight through people and have no knowledge of their presence unless we bumped into them. The sky would be transparent; our atmosphere does not absorb photons in the range of 16–33 centimetres. We would see straight out into our galaxy and the universe beyond. In this range of wavelengths, hydrogen glows brightly—it emits photons with a wavelength of 21 centimetres. Since most of the atoms in our universe are hydrogen, astronomers use radio telescopes tuned to detect these photons to survey and study galaxies. With our eyes we would be able to see the spectacular turbulent interstellar gaseous medium of our Milky Way galaxy stretching across the night sky.

Our world is one that we have quantified and described with axioms, rules and the laws of maths and physics. As we move to systems and scales that are unfamiliar, we lose our intuitive cognition—an important tool in understanding and thinking about the physical universe. Furthermore, our rules apply under certain conditions. The history of scientific discovery shows that, when we test our laws of physics in new regimes of nature, often some new physical mechanism becomes important. Consequently, we cannot describe the first instance in the history of our universe because we have not understood the rules and laws that link together gravity and quantum mechanics.

The Rules and Laws of Physics

Science began by observing recurring or reproducible effects. For example, every time you drop something, it falls downwards, never upwards. We do not usually have access to the entire chain of events that is responsible for this effect. We cannot see what is pulling the ball downwards when it falls, but we can see that the ball falls faster the longer it drops and we can use that observation to say something about the rules of gravity and motion. We try to understand the rules that make our mechanistic universe work, and one of the first things to be understood was the regularly repeating clockwork-like motions of the planets. Once we have rules that work on repeatable observations, we can use science to predict new phenomena and to design new useful or cool things.

It is, however, a difficult procedure. Rather like someone who does not know the rules of chess starting to watch a game in the middle and attempting to determine the rules of the game from the information present. The current placement of the pieces gives one set of clues, but it is not enough to guess how each piece may move. After observing the game play for a sufficiently long time, it is likely that you will be able to correctly state most of the rules. For example, you might never have seen the en passant move, but you might have been clever enough to predict that it could occur because of a certain configuration of the pieces and seeing how the game was played. This is similar to how antiparticles were discovered. They were first predicted from the rules and laws of nature, and their existence was ultimately verified through experiments.

Fortunately, for most purposes in our everyday lives, space, time and gravity all follow well defined "empirical" laws. Space seems to have the same properties in all places and directions. Time appears to march on at a relentless pace (as much as we would sometimes like to reverse time, we cannot), and gravity consistently pulls matter together over vast distances in our universe. If gravity, time and space were not predictable in this basic sense, stars would most likely not have formed, let alone led to the conditions whereby life on Earth could evolve. That everything in our universe obeys the laws of physics is good.

Underpinning this predictability and our well-behaved universe are the fundamental laws of conservation. A conservation law means that a particular measurable property of a system does not change as time proceeds. Let us take the law of conservation of energy. You cannot get something for nothing: Perpetual motion machines that endlessly do work would violate this most sacred principle of physics. What is energy? We know that electricity can cause things to move and turn, and that the Sun provides the energy for life on Earth. But what exactly is energy? It is an indirectly measured quantity that is often defined as the ability to do a certain amount of work, which essentially means changing the speed of something. You read earlier how the natural state of an object is to move at a constant velocity. Newton showed that to change the velocity of something one has to accelerate it by doing some work, by imparting a force.

Conservation of energy seems to make sense: After all, objects at rest do not spontaneously start moving. Ghosts and spirits violate the conservation laws, thus they cannot exist, at least not according to the physics that we know! So we have these laws of physics that seem to make very good sense, but where do they come from? Can we derive from first principles the fact that the energy or the speed of a moving object is conserved? The answer is yes we can, and it is rather surprising at first to learn that these laws follow from the symmetry properties of space and time. If space and time are continuous and well behaved, these laws drop straight out of the equations.

Emmy Noether was described by Einstein as the most important woman in the history of mathematics. Her early days epitomised the difficulties faced by women as academics in the 1900s. Noether was one of only two women to attend the University of Erlangen, and she had to obtain special permission from the lecturers to attend their classes. She taught at their mathematics institute for seven years without pay. It was not until after the First World War that attitudes changed and she was allowed to carry out her habilitation. Although she was still hired as an unpaid professor with limited duties and responsibilities. Noether made many outstanding contributions to science and is remembered in particular for a theorem named after her which shows how the fundamental conservation laws arise through the symmetries of nature.

Consider space. The speed at which you move does not depend on where you are in space: You can be at rest here, or over there. This symmetry of space leads to Newton's laws of motion. Space is also symmetric to rotations—we say it is "rotationally invariant". It does not matter how you orient a spinning object in space, its dynamics, or motion, are not affected by how it is positioned. This invariance to rotation gives us the conservation law of angular momentum or rotational motion. It is the reason that the Earth has continued to spin on its axis for over four billion years.

The relationship between symmetries and conservation laws is a beautiful unification of mathematics and fundamental physics. It also makes intuitive sense; the equations of an object moving through empty space do not include terms that depend on direction. It does not matter if that object is moving from right to left or left to right or up and down; it will move with the same behaviour. If there were a preferred direction, the symmetry would be broken and motion at a constant speed in the same direction would not occur.

Consider that time is symmetric and continuous—it can run forwards and backwards. As a consequence, all the laws of physics work forwards and backwards—they do not depend on the direction of time! Let us think about that for a minute. The symmetry of time gives us the most fundamental conservation law—that energy is conserved. That is rather profound! Does that mean that it is possible to travel backwards in time? Well, perhaps. I do not believe that you can jump to arbitrary places in time, since in my mind time is as continuous as space.

You cannot just step from one point to a more distant point without passing through all the intermediate steps.

Time

At the deepest level, we do not really understand what time is. Time is a measure of change, a measure of duration between events. We perceive that time proceeds only in one direction—forwards. Zeno also thought about time and gave us the "arrow paradox": "If everything when it occupies an equal space is at rest, and if that which is in locomotion is always occupying such a space at any moment, the flying arrow is therefore motionless." Zeno is stating that, at any instant in time, the arrow is frozen in space; and if time is composed of instants, motion is impossible since the arrow is always at rest. Even though Zeno's original writings have not survived the passage of time, Aristotle mentions and debates them at length in book six of his lengthy treatise on physics from 350 BC. You can find translations into English online.[1] Aristotle discussed whether or not space and time were infinitely divisible, whether space itself could be represented by a set of points if those points were in contact and so on. The Greeks were struggling to achieve a deeper understanding and were asking questions that we still ask today. There are several ways out of Zeno's paradoxes—can you think of one?

We are familiar with the passage of time using our clocks or watches; we can synchronise our time with a friend, and when we see them again our watches seem to agree. Unfortunately, basing our perception of time on our own experiences leads to an intuition that is not correct. If we compared times using very accurate watches with a friend who made a journey, the times would actually be different. Einstein showed that the time measured on a clock depends on the history of the clock and the journey that it has taken through space. He also showed that clocks run at different rates if they are in different gravitational fields—predictions that have been accurately verified numerous times. This leads to the undesirable consequence that no moment in time can be labelled universally because the present and events separated in space cannot be judged to occur simultaneously.

We have to consider time as a dimension, like another set of coordinate axes. Time and space, motion and gravitation are all connected through Einstein's equations of special and general relativity. Clocks run more slowly if they are moving faster. If you could travel at the speed of light, time stops completely. Consider a photon that was emitted shortly after the big bang. It may have travelled right across the universe for 13 billion years before being detected by our telescopes. At the speed of light, the photon would calculate that its journey is instantaneous. It is perhaps Einstein's theory of relativity that gives the best answer to the paradoxes of Zeno. Absolute space on its own does not exist since space and

---

[1] Physics, by Aristotle, 350 BC: http://classics.mit.edu/Aristotle/physics.html

time are intricately connected and cannot be separated. That does not stop me from questioning what the fabric of spacetime actually is. Is it something we can ever visualise, or can we only ever understand it from a geometrical and mathematical viewpoint?

We work with an enormous range of length and timescales. Using lasers, one can perform tomographic imaging of the motions of electrons with a precision of 10^{-17} seconds, which is the shortest interval of time that can be measured with current technology. We have no idea what strange things we might find if we were able to look on smaller length scales and smaller timescales; worlds within worlds and an infinite hierarchy of particles? Empty space? Or perhaps nothing that we can comprehend or measure? There is a similar uncertainty when we contemplate very large scales, beyond the horizon of our visible universe. Is our seemingly vast universe an irrelevant tiny patch within an infinite sea of other lonely universes? Or could these different regions and scales be connected by shortcuts in spacetime? Unfortunately, teleportation and wormholes remain in the realm of speculative research and science fiction, but they are not excluded per se.

So what is it that sets the direction of time? If you could instantly reverse the motions of all the particles in an isolated system, it would appear to change as if time were running backwards. In practise this is impossible for us to achieve. We cannot exactly reverse the motions of such a vast number of particles, and the concept of an isolated system is only possible as an imaginary thought experiment. Every particle of matter is affected gravitationally by every other particle in our visible universe. But imagine that we could exactly reverse the motions of all the particles in the universe. What would happen then? If we could observe such a universe without disturbing it, everything would appear to run backwards.

Time may be closely linked to entropy, which is a quantitative measure of flow and dispersal of energy. Liquid water has a much lower entropy than steam; as a gas the molecules have more freedom to occupy random positions. As entropy increases, the amount of energy available for useful work steadily decreases. The only physical law that has a one way direction is the second law of thermodynamics, which states that the entropy of an entire system can only increase. This relentless one-way increase in entropy has been linked with our perception of time moving in one direction.

The Austrian physicist Ludwig Boltzmann was one of the pioneers of the atomic theory of matter. He spent much time under attack from his colleagues, who in the late nineteenth century did not accept the existence of atoms. The equation that defines entropy is engraved on his gravestone in the central cemetery in Vienna. Some people say that it was the lack of acceptance of his theories that drove him to suicide in 1906. It was just a year later, following Einstein's doctoral thesis, that the atomic nature of matter was finally proven.

Boltzmann derived equations based on the motions of particles, which showed that entropy can only increase, thus giving a one-way direction of time. This leads us to an immediate paradox since a system that obeys molecular dynamics and the laws of physics is time-reversible. Therefore, one can reverse the dynamics and recover the initial state. So how does an irreversible process (the constant increase in entropy) come from time-symmetric physics? This question was asked by Johann Josef Loschmidt in 1876. The answer is that Boltzmann's derivation of the second law "assumes chaotic behaviour" in the way that atoms evolve over time through their collisions. This assumption has conceptually changed the system by introducing an element of time asymmetry. Chaos means that, in practice, a system cannot be exactly reversed since the slightest error or outside perturbation would change the entire system configuration at a later time. This sensitivity to small disturbances defines a chaotic system. It is often described using the example of the unsuspecting butterfly in South America that flaps its wings once and changes the weather in Europe. Chaos is the reason we cannot predict the future stock market or accurately model the weather more than a few days in advance.

The second law of thermodynamics is not a rigorous law of physics. It is a statement about finding a system in a certain "most likely" configuration and the most likely way the system will evolve in time. There is a finite probability that all the particles in the air in the room in which you are sitting can all be found in one corner, thus leaving your head in a vacuum! Sooner or later it will happen, and there is even a theorem formulated by the French scientist Henri Poincaré which implies that the air will occupy only one corner of the room an infinite number of times! Since you would have to stay in your room for much longer than the age of the universe for that configuration to happen once, I do not advise you to try this experiment at home. The take-home message here is that, in an average sense and for most of the time, the second law of thermodynamics is obeyed. The chaotic behaviour of such a large number of interacting particles in our universe is the reason that time cannot be made to run backwards.

Quantum Mechanics and Duality

In the seventeenth century, the Dutch physicist Christiaan Huygens (the inventor of the pendulum clock) believed that light must take the form of a wave. He came to this conclusion after studying how light behaved as it passed through different media such as glass and air. At the same time, Isaac Newton thought that light must be made of tiny particles since a focussed beam of light did not appear to behave like a water wave spreading out in all directions. However, experiments gave support for Huygens' theory, especially Thomas Young's famous double-slit experiment at the beginning of the nineteenth century.

My father and I were intrigued by this experiment, and we constructed our own version at home when I was 14. We purchased a fine double-slit, etched in dark glass, and I borrowed a strong light source from school. I remember setting up the apparatus in the attic and waiting until dark before focussing the light into a parallel beam illuminating the slits. The pattern of light that emerged was a series of alternating light and dark stripes which can be explained as constructive and destructive interference of "light waves" after they pass through the slits. (It is easy to reproduce the same experiment today simply by cutting two slits with a sharp knife about half a millimetre apart in thick paper and shining a laser pointer at the slits).

You can see a similar effect with water waves. Throw two stones into a still pond in different places and watch the ripples expand outwards. As the two sets of ripples meet, you can see a complex pattern forming. The ripples are waves that spread outwards because of the disturbance of the stones entering the water. However, the water molecules do not move outwards; they simply oscillate up and down, dragging their neighbours with them through their electromagnetic interactions. The visual impression is that of a wave moving across the surface of the water. When the ripples cross at a certain point, they can act together to form a larger wave or they can cancel each other out, leaving a flat surface. This is how Young interpreted the double-slit experiment. However, in 1921 Einstein received the Nobel Prize for demonstrating that light must in fact be treated as a quantized packet of energy, as a single particle, to explain the photoelectric effect. Things were getting confusing.

Until the 1960s the double-slit experiment was only performed with light; however, particles of matter were found to exhibit the same behaviour as light. When a beam of electrons was fired at two slits, they emerged on the other side producing the same pattern. The same wavelike behaviour is even seen when large molecules are used—most recently carbon fullerenes! However, the mind-blowing result occurs when single molecules are fired one at a time at the two slits and the positions are recorded as they emerge from the other side –the pattern is the same. Individual molecules that we can see with our microscopes somehow behave as a wave during their journey. The single molecule somehow passes through both slits and interferes with itself, which determines where it will land on the detector. Then, when the molecule does land on the detector, it triggers a response as if it were a single compact particle again. Light and matter behave simultaneously as waves and particles. If you close one slit, the dark-light pattern disappears. But how does a tiny molecule know that there are two slits there? This duality is at odds with logic. How can an object be black and white at the same time. How can it exist simultaneously in two states? This is only the beginning of a series of paradoxes and conceptual problems that occur in the quantum mechanical description of particles and light.

The paradoxes in quantum mechanics include the famous Schroedinger's cat thought experiment in which macroscopic objects, for example a cat, can be in two states simultaneously: both alive and dead. It is the observation or measurement of the system that "collapses the wave function" and puts the cat into a definitive alive or dead state. If that does not sound strange enough to you, consider the topic of quantum entanglement, which reveals rather spooky "action at a distance", or quantum effects travelling faster than the speed of light. By "entangling the wave functions" of pairs of particles—separating them by tens of kilometres and then observing the state of the particles, one can see that one particle is immediately influenced by the state of the other even though they are so far apart. Two observers in different locations can witness the same event simultaneously, although neither can control what the other sees.

Erwin Schrödinger joined the University of Zurich in 1921. In 1926 he published four famous papers which set out the wave formulation of quantum mechanics and derived his famous equation, which treats particles as a propagating probability wave. All of this was inspiration for the "Infinite Improbability Drive" in Douglas Adams' *Hitchhikers Guide to the Galaxy*, which was written in 1979. The probabilistic nature of quantum mechanics is inherent in Werner Heisenberg's uncertainty principle, which tells us that we can never know all of the properties of a particle at once. For example, we can measure the position of a particle with increasing accuracy but only at a corresponding decrease in the precision by which we can determine the particle's speed.

How can something be both a wave and a particle at the same time? It made no sense to me when I first discussed quantum mechanics with my father, and it makes little more sense to me today. But thanks to the likes of scientists such as Schrödinger, Richard Feynman, Wolfgang Pauli, Paul Dirac and many more, we now have a wonderful and detailed description of how particles behave and interact. Unfortunately, the deeper one delves into this topic, the more confused one's understanding usually becomes. Feynman writes on quantum mechanics in his lecture notes that the only thing that can be predicted is the probability of different events and concludes: "It may be a backward step, but no one has seen a way to avoid it." Schrödinger provided a beautiful mathematical description, a means of calculating how particles behave yet he himself wrote: "That it is an abstract, unintuitive mathematical construct is a scruple that almost always surfaces against new aids to thought and that carries no great message."

Forces

The advent of quantum mechanics enabled a deeper understanding of electromagnetism and the idea that the vacuum of space is actually filled with a sea of "virtual particles". Virtual particles appear in quantum electrodynamics, a theory that links together quantum mechanics, electromagnetism and special relativity. The mathematical formalism allows the interpretation of forces between particles as having resulted from an interaction with some lowest energy state of the vacuum—or virtual particles. They arise purely from the mathematics describing the situation and are a way to visualise what is going on. But their reality is open to debate. Virtual particles have never been seen, and theoretically they cannot be seen; so it would be reasonable to ask, Are they actually real?

In the classical picture, two similarly charged particles repel one another due to a constant exchange of energy and momentum through their combined electromagnetic fields. These fields propagate through all space from every charged particle, similar to the gravitational field. Particles will naturally follow certain trajectories in space, following field lines that minimise a certain combination of their energies. In the quantum picture, there are numerous exchanges of "virtual photons" between the two particles, and it is the effect of the emission and absorption of these events that changes the motions of the charged particles.

Virtual particles randomly appear and disappear on a timescale so short that they can travel less than one wavelength. Yet in that time the rules of quantum mechanics allow them to violate the laws of conservation of energy and momentum. The idea of cause and effect is quickly vanishing. Particles appearing from nowhere? Randomly? How? Why?

The electromagnetic forces that hold molecules together are only part of the reason that we can't walk through walls. To understand what makes objects 'solid' we also have to consider quantum degeneracy. That strange force that holds together neutron stars also maintains the integrity of atomic structures—it prevents atoms from overlapping and also stops the negatively charged electrons from falling into the positively charged nuclei.

Before you get the idea that this is crazy science fiction, I should point out that virtual particles make up the essence of quantum electrodynamics, the research area for which Feynman earned the Nobel Prize—a theory of how particles and forces work. It is the most well tested and experimentally validated theory we have—its predictions have been experimentally measured to be accurate to a precision of one part in a billion, which means getting the ninth decimal place correct!

Our understanding, or lack of it, is illustrated by the fact that we cannot repeatedly ask Why? or How? more than a couple of times before we reveal our ignorance.

Why did the apple fall?

Newton had the foresight to ask this question and the intelligence to find a solution. He wondered if the same mysterious attraction that the Earth had for the apple could reach to the Moon and beyond and if so, why the Moon did not fall to Earth like the apple.

Newton would have answered, Because all massive objects attract each other through the force of gravity.

Why does gravity attract objects?

Newton could not answer this question, but Einstein gave one possible explanation. In his theory of general relativity gravity is a result of the curvature of spacetime caused by the presence of matter. The starting point for general relativity is that there is no difference between the gravitational acceleration and the kinematic acceleration. This is called the equivalence principle. If you happen to be inside an elevator that is falling freely because the cable snapped, you would be weightless since weight is the force exerted on you by the Earth. You would be literally floating in the air, albeit for a very brief period before you hit the ground. There is no way that you could distinguish that period of weightlessness from really floating in space away from any massive object. Thus, acceleration is equivalent to gravitation. In general relativity, objects move along locally straight paths in curved spacetime such that they "feel" acceleration equivalent to the gravitational force of the matter.

How does matter curve spacetime?

Einstein could not answer this question. No one can, yet. On the one hand, Einstein's theory of gravitation works perfectly well as far as we humans are concerned. There is no need for a better theory—it is tested and accurate to a precision beyond what we should ever need for our day-to-day lives. On the other hand, it is still frustrating not to be able to understand the answer one level deeper. To answer this question may require that theory of quantum gravity, which could be similar to quantum field theory in which the electromagnetic, strong and weak forces are mediated by bosons. It is possible that a new class of virtual particles called gravitons may be filling the void between all particles of matter and may be responsible for the force of gravity. Of course, even after we refine our theories of gravity, there is still the possibility that we could continue to ask why a few more times.

A Deterministic Universe

A digital computer is completely deterministic. If you run the same program twice, you will get exactly the same answer. The computer does exactly what we instruct it to do, step by step, one instruction following another. As we try to understand the purpose to life, to our universe, at some point we come to the question of whether or not our consciousness and our decisions are also deterministic. Is there free will such that we can change the direction and outcome of our lives? Or is each thought and action we make based on a previous event?

The laws of Newton are deterministic: Every action has an equal and opposite reaction. If everything in our universe obeyed Newton's laws, then everything that happened in our universe would be predetermined and, in principle, could be predicted. Perhaps even our thoughts would follow a deterministic path, in which case we could argue that we are not responsible for our actions since there is no free will. A deterministic universe is one in which the path of every atom and every photon can be calculated given the configuration at any one instant in time. In practice, it is impossible to predict the outcome of events of even the simplest system because of chaos, since the motion of every particle in the universe affects every other particle in the universe. But the question of whether or not everything is a result of cause and effect is still an interesting one to ask from a philosophical viewpoint.

If any physical process in our universe were to occur randomly, we would lose predictability—the future would always remain unknown. It would seem strange that nature could have an implicit randomness, the equivalent of dice that "it" throws to determine an outcome. Einstein remarked that god does not play dice—a prejudice that naturally arises as one studies classical physics and then tries to understand quantum physics. Einstein was referring poetically to "nature": If this was all designed by a god, then god would be clever enough to know the outcome of the big experimental laboratory, the universe. Our "designer" would know exactly where each atom would go, where intelligent life would form in the universe and how we would all be obsessed with the Internet on a small typical rocky planet orbiting a normal star within a galaxy that looks like billions of others! What would be the point in creating this universe? As events unfold in the distant future, we will find that the universe has become a very inhospitable place for life. We certainly cannot be an important component of the universe in the eyes of a creator since it appears that, as the universe ages, the prospects and ability for life to thrive steadily decrease.

The standard interpretations of quantum mechanics rely on the principle that matter on the smallest scales behaves in certain ways to which we are not accustomed. You cannot measure things to arbitrary accuracy since particles are fuzzy and cannot be localised—we could not even measure the initial conditions accurately enough to calculate the future. If this is true, the universe is not deterministic. The accuracy with which we can predict the state of a system at a future time rapidly diminishes because of unforeseen events. However, this apparent unpredictability may only be our interpretation of quantum mechanics and not the reality. A phenomenon can be interpreted in different ways yet still be consistent with the equations and the observations. One can take a working theory and reformulate it in such a way that all the same results are recovered; it is only the physical interpretation that is different.

As an example, our standard theories of electromagnetism are based on the idea that electromagnetic waves carry information in one direction in time. In 1945 Feynman and John Archibald Wheeler reformulated the theory by introducing the postulate that information can simultaneously travel backwards and forwards in time. Since the laws of physics are all time-reversible, they considered a time-symmetric approach to the theory which they illustrated using the example of the emission and detection of a photon (an electromagnetic wave). The process of emission produces a wave of half the amplitude travelling forwards in time and another half-amplitude wave travelling backwards in time. This can be thought of as an interchange that occurs between the emitter and the detector—a sort of transaction-verification process. This reformulation leads to the same equations but has a rather different physical interpretation.

Quantum mechanics can also be recast into a similar time-symmetric form. The consequences are remarkable. Events in the future influence events in the past through simultaneously propagating waves passing forwards and backwards in time. (In the language of more advanced physics, this means using a Lagrangian rather than a Hamiltonian approach—they are equivalent and the latter is derived from the former, but they differ fundamentally in their interpretations. The Hamiltonian advances a system to a new time using the position and momentum at a previous time. The Lagrangian depends only on the position now and the position at a later time.)

With this interpretation of quantum mechanics, the destiny of our universe is already set. Events in the future reach backwards in time and conspire with events in the past to make the present a reality. The probabilistic fuzziness that appears to be present in nature can be thought of as a barrier that prevents single events in the future from affecting events in the past and thus changing the future. Despite this fuzzy quantum barrier, in this interpretation of quantum mechanics, the past and future of the universe is theoretically predictable. This reformulation of quantum mechanics by Israeli physicist Yakir Aharonov in 1962 is completely deterministic. This has the philosophical implication that there is no free will. Every thought and action we take follows as a direct consequence of a previous event—thus your path through life is already predetermined.

Our understanding of the universe does not necessarily rest on firm ground since it is not complete. In fact, in 1931 the Austrian mathematician and philosopher Kurt Gödel showed that a mathematical understanding of most systems can never be complete. But still, it can be better. Given the rate of advancement of knowledge and from all of our past experiences, there are bound to be surprises in our understanding. After all, it is not as though we live in a preferential time where we have made all the big discoveries that are possible. Far from it—I am just disappointed that I will not see how science progresses over the next two thousand years.

Chapter 11
The Final Chapter

B. Moore, *Elephants in Space*, Astronomers' Universe,
DOI 10.1007/978-3-319-05672-2_11, © Springer International Publishing Switzerland 2014

I was invited to join the University of Zurich to begin a new research activity in astrophysics and cosmology. To build a successful group, you need to hire the best people, to have good ideas and to create an environment that is both stimulating and fun. The late night "work venue" became the El Lokal bar, which has an atmosphere conducive for creative thinking and good discussions. A giant skeleton hangs from the ceiling, and the walls are covered with images of interesting personalities such as Che Guevara and Frank Zappa. We wanted our own supercomputer so that we could simulate the universe to study how stars, galaxies and planets emerge from the smooth and hot big bang. The problem was that in 2002 commercial supercomputers were all very expensive. Joachim explained how he believed that we could build our own, and he began to sketch out his idea for the "zBox". We talked until the early hours, refining the idea and discussing the cool new research that such a machine would allow us to carry out. We would buy 144 motherboards that could each hold four CPUs—commodity parts that would normally be used in a home PC, but we would connect them all together with a high-speed communications network. This would allow us to make calculations in one month that would take 50 years on the fastest home computer. The main challenge was to design a machine that could fit in an office and did not overheat. Each motherboard takes about half a kilowatt of power—that's 70,000 watts of electricity coming into the machine, of which almost 100 percent comes out as waste heat. That is like having 70 electric heaters in a room the size of a bedroom. We had a tight budget and limited space, but those constraints forced creativity. Joachim's design was novel. All the components would be mounted onto shelves that could be pulled in and out from each side of a custom-made cube-shaped rack. We would inject cold air into the centre of the machine which would flow out across the motherboards, keeping them chilled. Encouraged by bribes of beer and pizza, the first PhD students we hired helped assemble the machine. After the construction was finished and the power turned on, we were all overjoyed to see 144 blinking green lights signalling the signs of life. We had managed to construct one of the world's highest-density supercomputers and the fastest computer in Switzerland at less than a third of the cost of machines provided by the big commercial companies.

We have come a long way already. "Must we go further?" you might ask. Well, I cannot resist. What lessons can be learned from our evolutionary path and the history of our knowledge? Eschatology is the study of the ultimate destiny of humanity. Previously in the realm of religion and speculative philosophy, towards the end of the last century several prominent scientists began to pursue the topic according to what we really know about the universe. I think you can come to your own conclusions. But before I give you mine, let us first look first at the ultimate destiny of our galaxy. After all, intelligent life must co-exist and survive together with matter and energy in some form.

Ten trillion years after the big bang all the stars in our galaxy have stopped shining, and no new stars are forming. The night sky is dark in all directions and in all places. At this time the galaxy bears no resemblance to its bright colourful state today. Most of the stars we currently see have ended their lives as white dwarfs, and these cold dark objects continue to orbit through the galaxy. A hundred billion stellar remnants slowly cool down like the embers of a fire, fading from sight as they radiate their energy away into space. After one thousand trillion years, the white dwarfs will become icy cold—black dwarfs. They have cooled down to a temperature less than −200 degrees centigrade, and they glow dimly with an effective luminosity that is one trillionth of that of the Sun today. If you were to sum up all the residual radiation from these remnants of stars, it would come to less energy than one hydrogen-burning star. No black dwarfs have been discovered in our galaxy. But at this point we do not expect to find any since there has not been enough time for the white dwarfs to cool down to such a state.

Brown dwarfs, those failed stars that never became massive enough to trigger nuclear fusion, are by now frozen spheres of solid hydrogen. Any residual gaseous interstellar medium has long since been accreted by the supermassive black hole at the centre of the galaxy. The most massive stars ended their lives as spectacular supernovae a long time ago, their centres collapsing into black holes or neutron stars. They are a minority, yet there are at least a billion of these exotic remnants drifting throughout our dark galaxy. The hundred billion distant galaxies have also faded from sight as their stars burn out and the expansion of space makes them invisible to us. Ten trillion years after the big bang, we see nothing beyond the galaxy but empty darkness.

The End of Our Galaxy

As we lose our contact with everything beyond the galaxy, what about the galaxy itself? It is the remarkable attractive force of gravity that caused structure to appear from the nearly smooth initial conditions that emerged from the big bang. Stars shine and our galaxy formed thanks to gravity pulling material together. Yet it is also gravity that will be responsible for its ultimate fate in the far, far future. What happens to our galaxy on even longer timescales? Astrophysicists Fred Adams and Gregory Laughlin worked out many of the details in a lengthy scientific paper published in 1996 titled “A dying universe: the long-term fate and evolution of astrophysical objects”. Ten trillion years after the big bang, our galaxy is a nearly spherical collection of orbiting dense stellar remnants, failed stars, black holes, neutron stars and icy cold planets. It all sounds rather ominous, but there are still some interesting events to come as the galaxy slowly restructures itself as it grows colder and colder. As the remains of dead stars and brown dwarfs orbit in darkness, they sometimes pass rather close to each other, like large ships in the night, feeling each other’s presence through their mutual gravitational attraction.

As time passes, gravity continues its relentless goal of achieving a maximum entropy state and tries to achieve equipartition among its components. Equipartition means that any set of gravitating objects, like stars, will slowly redistribute their orbital energies such that they all have the same energy. This occurs through those gravitational encounters between the stellar remnants which tend to cause the more massive objects to slow down while the lighter objects speed up. As a consequence of all of these encounters, the most massive objects sink deeper into the centre of the galaxy, passing closer and closer to its central supermassive black hole. There is no equilibrium state for a gravitating system. The central regions of the galaxy become denser and “hotter” as the stellar remnants sink, since as they do they gain gravitational energy and move faster. Eventually, they are consumed and disappear within the event horizon of the black hole, which as a result grows slowly in mass.

As the outer, lighter objects also gain kinetic energy, they begin to orbit to larger and larger distances until eventually they escape the gravitational attraction of our galaxy altogether. They disappear into the dark and empty universe and are finally taken away from sight by the incessant expansion of space.

The process is rather like the way that water evaporates from your skin. The water molecules have a range of speeds, and the fastest-moving ones have enough energy to escape from its surface into the air. This leaves behind the colder-moving molecules, thus cooling you down. That is one reason why you sweat. To escape the gravitational attraction of our galaxy, objects need to reach a speed that is greater than about 500 kilometres per second. Each gravitational encounter changes the velocity of a star by a tiny amount so it takes thousands of trillions of years to reach the escape velocity, but eventually it does. No one has calculated in detail how the galaxy evolves on the timescales that it does as it is restructured by gravity. However, "back–of-the-envelope" calculations of these effects suggest that, on a timescale of about 10^{25} years, our galaxy slowly disappears, leaving behind a single supermassive black hole that will have eaten its fill and grown to a monstrous size, ten billion times the mass of the Sun. Everything else that was in our Galaxy is now beyond its detectable horizon. I am still writing in powers of 10, but the size of this number should be appreciated. Our galaxy disappears over an amount of time that is a thousand trillion times longer than the time for which the universe has existed. Finally, the role of gravity in the universe is just about over. That is the uniqueness of gravitating systems—there is no maximum entropy or equilibrium state. It will keep evolving until everything is thrown away into space, leaving a single object behind. Even its dark matter halo, made of about 10^{50} tiny particles, will eventually evaporate away.

All that is left of the galaxy in the distant future will be a single massive black hole. Everything else will be flying radially away for eternity. This is a bizarre reversal of the events by which the galaxy assembled and came together. Over this enormous timescale it seems as though life can flourish only in a very brief window of time. Without the heat energy from stars it is unlikely that life would evolve again from a molecular level. The time interval over which the galaxy can evolve and abundantly host life as we know it spans from 10^9 to 10^{14} years after the big bang. Our own civilisation emerged in the first 0.1 percent of this time period. Beyond 10^{14} years from now, life can survive only if it acquires and maintains energy, using ever-dwindling supplies. Such an achievement would require technological advancements that are beyond even the dreams of most science fiction writers. But perhaps it is possible.

In fact, the galaxy is not completely dark over this enormous time period during which it slowly disperses. Brief moments of light occur as rare random collisions take place between failed stars or the white dwarf remnants of stars. When two brown dwarfs collide directly, the combined mass can be sufficient to generate temperatures high enough to ignite and sustain hydrogen fusion. The resulting object is a red dwarf star, alone in the galaxy and shining for a trillion years. This leaves the incredible possibility of new life appearing even in our galaxy, evolving and forming on the rocky planets that could emerge from the collisions or whose habitats become defrosted by the new star.

Imagine that—the prospect of life emerging from the darkness to witness the end of the galaxy. Even if that life evolved to an advanced stage capable of constructing sensitive telescopes, it would not be able to ascertain the history of the universe. It would not see evidence of the big bang or an expanding universe since the distant galaxies would have faded out of view. It would be extremely difficult to detect anything beyond its dim star. All the remaining objects in the galaxy would glow with less than the power of a candle held a light year away. It would be extremely difficult even to detect its own motion through space or to know that it lived in a galaxy with billions of dark objects orbiting around. To life in the distant future, the universe would consist of its single faint star with no knowledge or ability to ever understand how that star got there!

Even rarer would be collisions between the carbon- and oxygen-rich white dwarfs that yield a remnant with a mass larger than 1.4 times the mass of the Sun. The resulting object would immediately explode as a supernova, lighting up the galaxy for several weeks; however, the timescale between these events is immense.

So far, everything I have mentioned is probably really going to happen. It is not made up or based on complete speculation but has scientific grounding. Of course, there is always a possibility that some new discovery will radically change our understanding. That discovery is not likely to change our general picture of the big bang or the history of the universe, but it could impact on our distant future. Our predictive powers for the long-term future of the universe rest on certain assumptions. We assume that the laws of physics do not change with time, which we know has been true in the past by studying chemical reactions that took place billions of years ago. We also assume that all physics is known—an assumption that has proved incorrect repeatedly in the past.

Prospects for Life

It seems crazy to contemplate such long periods of time that seem irrelevant to our everyday lives. But time is relative. Some species of mayflies only live for about an hour, just enough time to find a partner, mate and then die—some without ever even eating. They have been doing that for several hundred million years and have barely evolved since they appeared among the first things to fly on our planet. Most of us live a million times longer than the shortest-lived mayflies; we must seem unimaginably old to these creatures. With that thought, contemplate a creature that lived for a million years longer than an average human. Some species of sponges and certain trees can live for over 10,000 years. But longevity does not seem to be a goal of evolution. You might think that a short life cycle would lead to more rapid evolution since there would be more cycles within which change could occur, but mayflies and ants show this not to be the case. Evolution seems to work towards finding equilibrium, a niche where a stable population can simply exist.

Ant colonies are an example of complex behaviour that is more than just the sum of its parts, yet they have no grand purpose. Each individual ant has a small function in the role of the colony; but working as a unit, the ant colony is a city of life that functions in a remarkably complex way. The ant species is very successful. It has reached its optimum state and has barely changed or evolved over the past millions of years. Ants do not appear to question purpose or meaning. Individualism is not important—the ants strive only to maintain the continuation of their queen.

We humans have reached a strange stage whereby our evolution is artificially changed through intervention not seen before in nature. We have understood the conditions needed to promote longer lifespans: healthy lifestyle and diet. Through medical intervention we can keep people alive who would have quickly perished in the past. However, many biologists think that it is unlikely that the average life expectancy of humans can ever exceed much more than one hundred years. In fact, it is not known why we die of "old age", and there are many theories, from limitations on cellular reproduction to genetic encoding.

How can we enable longevity? Life does not necessarily need carbon-based molecules and complex DNA-based cellular structures to exist. After all, it is those structures that are susceptible to the diseases that eventually lead to the malfunctioning of our bodies. Life could potentially exist in very different forms. In 1957 the distinguished British astrophysicist Fred Hoyle wrote a science fiction book called *The Black Cloud* in which an immense cloud of interstellar gas engulfs the Sun and sends the Earth's climate out of control in what was in effect a nuclear winter. The cloud is hyper-intelligent, thinking via electromagnetic signals like a giant computer. Unlikely, yes. Impossible, no.

If it seems unlikely that intelligent thinking life could develop in a completely different medium, consider whether it could transition to that stage after a carbon-based evolutionary route. *R.U.R.* is a 1920 science fiction play by Czech writer Karel Čapek. R.U.R. stands for Rossum's Universal Robots, the first time the word *robots* was introduced to the English language. His play depicts clonelike humans that can "think for themselves", but eventually a hostile robot rebellion leads to the near extinction of the human race. It is to avoid scenarios such as this that the prolific and far-thinking science fiction author Isaac Asimov proposed his three laws of robotics, intended to secure the safety and welfare of humans among robots. But even hardware-encoded rules can create an atmosphere of false security since rules based on language are fraught with loopholes and philosophical paradoxes. It is possible, and some argue it is likely, that machines will eventually take over as the dominant "species".

The exponential increase in technology is both an incredible and frightening phenomenon. Exponential growth occurs when the size of something in the future is proportional to its size today such that the rate of growth is constantly increasing. For example, the world's population, the economy, and computer speeds are all increasing exponentially fast. Alan Turing and John von Neumann both suggested that, at some point in our future, a "technological singularity" will occur that will mark a transition between the time that humans and machines dominate intellectually. The term *singularity* is used because it is impossible to predict what would happen after the development of an artificial superintelligence and its ability to recursively redesign improved versions of itself. It is not the same as the singularity that occurs when matter collapses into a black hole, but it has the same connotation. It marks a point beyond which we lose predictability because our understanding fails. Several authors have predicted that the technological singularity will occur as soon as mid-way into this twenty-first century.

I have already discussed the similarity between our brains and our digital computers. We interface with the world through our senses, which are a complex network of neurons and nerves that function by means of electrochemical reactions. Can we replace that interface with artificial signals that make us think we are interfacing with the world, whereas in fact our consciousness is inside a machine? Can we simply create a fake reality by coupling our senses to electrochemical sensors connected to a computer? There are plenty of books and movies based on this theme. Zeno's idea that movement is only a perception could become a reality. But you know that you are not embedded in a virtual universe created by a computer. Don't you?!

If we could interface our brains with our machines and maintain control over the combination, we would have a possibility of maintaining a conscious state for as long as we have the energy to make our machines function. This is another of those ideas in the realm of science fiction, yet it remains a distinct possibility. Our "brain waves" (electrical signals picked up on the surface of our skin, around the scalp, that result from the synchronised firing of millions of neurons) can already be captured and used to control simple remote actions. It is not so far-fetched that, in the future, we will be able to interface efficiently with machines or with other conscious beings.

We seek excitement, thrills and pleasure. These are a result of chemical reactions and impulses that flow through our nervous system to our brains. There is no reason why these inputs cannot be simulated and reproduced artificially to such an extent that we never have to leave our beds. Over the past 20 years alone the degree of reality in computer games has progressed enormously, from ascii-based animations to full three-dimensional virtual reality. It is this demand for realism that has continued the development of ever-faster computers and graphics cards. These components were not designed and developed with science in mind, but we can take advantage of them since they are at the heart of most of our scientific supercomputers. I look forward to seeing what the next 20 years of gaming technology brings and can only dream of our virtual environments 200 years from now.

As our physical bodies decay or become unnecessary, our mind and thoughts and feelings could continue to exist. It is theoretically possible that life could continue to survive for as long as there are energy sources available. To make a calculation inside our computer requires energy for something to change its state. Likewise, for our brain to recall a memory or to make a decision also requires energy. However, there is a large resource of energy in our galaxy to enable life to continue for a long, long time. It would be possible for a highly evolved galactic civilisation to find the failed stars—the brown dwarfs—and to manipulate their orbits, storing them in special stable orbital configurations so that they can be used for energy when needed. Perhaps the matter content can be placed into fusion machines to generate energy, or they could be deliberately crashed together, igniting a new star that can shine for another trillion years.

In his book on the birth of the universe, *The First Three Minutes,* Steven Weinberg states, "The more the universe seems comprehensible, the more it also seems pointless." He is pessimistic about the long-term future of life in the universe and believes that it faces a death by heat if it collapses and a death by coldness if it were to expand forever. This inspired the British theoretical physicist Freeman Dyson to write a scientific paper in 1979 on the possibility of eternal life in an infinite open universe, one of the first serious eschatological studies.

It is remarkable how our universe has evolved since the big bang, passing through a phase that enabled the conditions for life to evolve. As time ticks on into the future, it becomes more and more difficult for life as we know it to function. Dyson devised a clever way in which an intelligent being could think an infinite number of thoughts in an open universe using a finite amount of energy. The intelligent beings would begin by storing a certain amount of energy, and they would then use some fraction, say one-half, of this energy to power their thoughts. When the energy gradient created by unleashing this fraction of the stored fuel was exhausted, the beings would enter a state of zero-energy-consumption "hibernation" as the universe cooled around them. Once the universe had cooled sufficiently, an "alarm clock" would wake up the life. Half of the remaining half (one quarter of the original energy) of the intelligent beings' fuel reserves would once again be released, powering a brief period of thought once more. This would continue, with smaller and smaller amounts of energy being released. As time ticked on, the time between thoughts would be longer and longer, but there would still be an infinite number of them.

One problem with this is that to have (and keep) new thoughts, one needs to be able to store them, and we may only have a finite amount of storage material to work with. To have new thoughts is not very appealing if we have to erase old thoughts in order to store the new. As Dyson states, "To be immortal with a finite memory is highly unsatisfactory; it seems hardly worthwhile to be immortal if one must ultimately erase all trace of one's origins in order to make room for new experience." Dyson goes on to argue that memories can be stored digitally or in analogue form. While a finite material resource limits digital storage, there is no limit to the capacity of an analogue memory constructed from a finite number of components. For example, the angle between two gyroscopes spinning in space could be used to store information, and the capacity of this unit is equal to the number of digits to which the angle can be measured. The angle can be set exactly equal to a string of numbers, and those numbers can represent a date, or an image, just like we store information in a memory stick. Unfortunately, thanks to quantum mechanics and the expansion of space, angles and distances cannot be measured to infinite accuracy. Consequently, we are left with the prospects of a very large but finite memory.

In another research paper on the topic of the future of humanity published in 2000, Lawrence Krauss and Glenn Starkman revisited Dyson's arguments in light of finite energy resources, quantum mechanics and dark energy. They discussed in more detail the prospects for life to continue. The outlook is indeed bleak, although in principle enough energy is stored in the galaxy to maintain life forms with metabolisms equivalent to our own, for at least as long as the timescale over which the galaxy restructures itself and evaporates, and perhaps much longer. However, Krauss and Starkman argue that the hibernation strategy is doomed because the alarm clocks that wake the intelligent beings will become unreliable and ultimately fail thanks to quantum mechanical effects. They conclude, "Assuming that consciousness has a physical computational basis, and therefore is ultimately governed by quantum mechanics, life cannot be eternal."

With finite matter and energy reserves, it seems rather difficult for life to continue forever, although it can exist for a long time. Perhaps energy could be extracted from the remaining massive black hole that contains the matter of billions of stars. In 1969, the British mathematician and theoretical physicist Roger Penrose devised a mechanism to extract a large fraction of rotational energy from black holes. It is rather elaborate and requires sending matter at just the right trajectory so that it skirts the event horizon—part of the matter can emerge with a lot more energy than it started with, and that energy could be captured and used for a very long time, but not for eternity. Moreover, there is likely a final nail in the coffin for the long-term existence of life.

The End of Matter Itself

Ultimately, whether life can exist for an eternity rests on the question of whether its basic constituents, the protons and neutrons, are stable. We know that a free neutron is unstable and will decay into a proton by emitting an electron and a neutrino. The half-life for this process is about 15 minutes. However, once the neutrons are bound inside the nucleus, they are stable thanks to the presence of the protons. But what about the protons—can they last forever? If atomic nuclei and, in particular, protons were unstable, the material base for intelligent-thinking beings could eventually disappear since all matter could disappear, leaving behind only photons.

But nothing lasts forever, and even protons are expected to decay. This expectation is a prediction from "grand unified theories" which attempt to create a single theory that describes the electromagnetic and the strong and weak interactions that occur in nature. This is the first step for constructing a grand "theory of everything" that also includes the gravitational interaction. A grand unified theory would be able to explain why the electrical charge of a proton is apparently identical to that of an electron and why the universe ended up with an excess of matter over antimatter. This is profound stuff indeed. The energy scale at which these three forces are unified is just above the Planck scale and far above that which the LHC can probe, or in fact any possible particle accelerator we could build on Earth. However, one of the predictions of these theories is that the proton decays into elementary particles on a timescale that may be about 10^{36} years. That is an incomprehensibly long timescale, yet experiments already set constraints on the lifetime of a proton that is not far from this.

Protons are expected to decay into two photons and a positron, the antipartner of the electron. The positron would quickly encounter an electron, and these would annihilate each other and form two more photons. The entire mass of the protons eventually turns into pure energy in the form of these four high-energy photons. The Super-Kamiokande experiment has recently set a lower limit on the lifetime of a proton that decays in this way—a lifetime that is longer than 10^{34} years.

It is quite an experiment: A giant tank of 50,000 tonnes of ultrapure water surrounded by 13,000 sensitive photon detectors that stare into the darkness at the water. It is located a kilometre underground in the Mozumi zinc mine under Mount Kamioka in Japan. One purpose of this vast detector is to spot neutrinos by searching for those rare cases in which a neutrino encounters an atomic nucleus and produces a charged particle that can be detected via the production of photons. In this way, the Super-Kamiokande detector can actually observe neutrinos that are produced in the fusion reactions that occur deep inside the Sun.

While it can take a million years for the photons produced at the centre of the Sun to reach the Earth, the neutrinos escape straight away and reach the Earth eight minutes later. A trillion neutrinos produced inside the Sun pass through your body every second. Of those neutrinos, a single one is expected to be stopped somewhere inside the Earth by colliding with an atomic nucleus. Now and again, a neutrino will pass close enough to the nucleus of a hydrogen or oxygen atom in the Kamiokande water tank, creating a tiny visible burst of light. The direction from which the neutrino came can be measured by the photon detectors, thus allowing an image of the Sun in neutrinos to be constructed! These observations confirm that the nuclear fusion engine of our Sun is working just fine. If the Sun "switched off" tomorrow, we would know this since the neutrino flux would stop. That would be an early warning system for the end of our main energy source.

The experiment is also designed to search for proton decays because the giant tank of water contains about 10^{34} protons and the photon counters can detect the decay of just one of them. None have been observed to decay, thus the experiment has set a lower limit to its half-life. To provide stronger constraints, we either need to build larger detectors or keep searching for a longer interval of time. Detecting proton decay would be remarkable and provide verification that a grand unified theory exists. It would also provide the timescale for the survival of life based on particles—the ultimate end of life and our universe.

If the proton half-life is 10^{36} years, then half of the remaining mass of whatever atoms are left in our galaxy would have vanished by this time. In 10^{39} years the remaining number of protons would have had their numbers halved a thousand times over, which reduces the remaining total mass to nothing. Nothing lasts forever, not even diamonds. If the proton lifetime is of the length of time predicted by grand unified theories, after about 10^{39} years all the protons will have decayed, and all that will exist in our galaxy is its single massive black hole. In 1974 the British theoretical physicist Stephen Hawking showed that even black holes do not last forever. The lonely supermassive black hole that was once our galaxy will eventually decay into photons and other relativistic particles. The virtual particles created near the event horizon sometimes find themselves separated, with one inside the black hole and one outside, having escaped. The entire mass of our black hole slowly evaporates into space on a timescale of a lengthy 10^{60} years.

And that's it. By this time, not a trace of our galaxy exists.

Final Thoughts

Three thousand years ago, a few far-thinking ancient Greeks started to question the world around them in a scientific way. They observed, experimented and speculated. Their search for knowledge of the physical world around us, of our origins and place in the universe, fuelled our advancement. It was these few deep-thinking and original individuals that produced the understanding necessary for us to progress. When Anaximander thought about understanding the nature of air, it was just a curiosity; it had no practical value. But if we had not developed the theory and eventual means to prove the existence of atoms, our world would be a very different place. Without the knowledge of how atoms and molecules behave, we would not be able to manipulate them to form the complex materials and chemicals that we now take for granted. Without questioning minds and the desire to understand, we would literally still be in the Dark Ages.

I have tried to convey the vast amount of knowledge that we have accumulated over this short time period. I hope to have demonstrated the rigorous nature and importance of our scientific methodology. I have tried to portray how incredible our universe is and how the reality is sometimes more remarkable than our imagination. I have also attempted to show some of the limitations of our understanding. Perhaps science can never answer the ultimate questions of how and why our universe came into existence and why it is the way it is. Perhaps we are scratching the surface of all there is to know, both on the smallest and on the largest scales.

The universe is ours to explore. I have argued, as others, that this is both possible and necessary for our continued existence as a species. But at what cost? The LHC cost 10 billion dollars. That sounds like a substantial amount of money. But investing in research and development to answer such fundamental questions is essential for the continued advancement of the human race. The technology we have today is a consequence of investing in basic research, including areas that have no obvious immediate financial payback. Who could have predicted that the World Wide Web would result from research carried out by the physicists at CERN in 1989?

To put this into perspective, in 2013, 150 countries across the entire planet spent a grand total of over two trillion dollars on defence. The total amount of money spent on space missions is only about one percent of this amount: The yearly investment that has funded the LHC is just 0.05 percent. Whereas the United States and China lead the arms race, even small countries like Switzerland, with no obvious need for a military force, spend several billion dollars a year on weapons and training. To what purpose? To protect against the same undesirable human tendencies that have always been present within our species. What a shameful situation for a supposedly intelligent species.

In 1903 the British philosopher Bertrand Russell wrote the following passage in *A Free Man's Worship*: "That Man is the product of causes which had no prevision of the end they were achieving; that his origin, his growth, his hopes and fears, his loves and his beliefs, are but the outcome of accidental collocations of atoms; that no fire, no heroism, no intensity of thought and feeling, can preserve an individual life beyond the grave; that all the labours of the ages, all the devotion, all the inspiration, all the noonday brightness of human genius, are destined to extinction in the vast death of the solar system, and that the whole temple of Man's achievement must inevitably be buried beneath the debris of a universe in ruins—all these things, if not quite beyond dispute, are yet so nearly certain, that no philosophy which rejects them can hope to stand. Only within the scaffolding of these truths, only on the firm foundation of unyielding despair, can the soul's habitation be safely built." These are insightful thoughts prior to the realisation of the detailed workings of stars, the scale of our universe or the big bang. That however powerful humans may be, in many and varied ways, physical or emotional, they still cannot control what will become of the universe.

Of all the great philosophical texts, books and stories, the most printed of all time is the Bible, with several billion copies in print in over one thousand languages and dialects. Perhaps the Bible would not have become such a bestseller if it had started "And god thought it would be a good idea to create a rapidly expanding quark-gluon plasma." But it is a revealing fact that the Bible does not contain any of the knowledge about the origin of the universe or the workings of the solar system, or any of the scientific understanding that we have gained over the past three thousand years. There are no clues or insights into the workings of our natural world. And the same is true of all the ancient religious texts that are believed and followed by so many.

Genesis, the story of creation in the Old Testament, is based on Babylonian myths. The gospels that recount the life of Jesus were written several generations after he had died, based on stories passed on through word of mouth. That some people can take these stories as literal truth is beyond my comprehension. If our universe were created by an intelligent designer, then that designer did not give the emergence of life much thought since life can only exist for a narrow window of time within our universe.

Our consciousness is believed by many to be a "soul" that exists separately from our physical bodies. I think of our consciousness as no more than the workings of the elaborate chemical computer which we call our brain. When we die, the memories and thoughts imprinted into our neural network can no longer be accessed. Once the energy supply ends, the molecules break down and our bodies and brain decompose eventually into dust. It is no longer possible to reach the information and memories we once stored and cherished; they die with us. The happiness we enjoyed, the pain we suffered and the love we felt are all lost.

People grieve and are sad when they lose a loved one. But that is often a selfish grief, a mourning of personal loss; after all, it is all over for the other, who cannot feel or suffer anymore. Many people are scared of death. Sometimes I wish I could believe that there is more to all of this, that death is just a step in our existence and that our consciousness lives on, either in heaven or hell. Unfortunately, I give very little thought to life after death because I believe that there is none. For a hesitant moment I could be envious of those people who believe in a continuation, an existence beyond our mortal molecular host. But we should not live our lives according to false dreams and fallacious hopes.

Most of us are quickly forgotten when we die. Knowledge of our past is soon lost. How many of you know the names and nationalities of your great, great grandparents? That is just three generations. What about going back ten generations to the eighteenth century? That is over a thousand ancestors. The record perhaps goes to Confucius, the Chinese philosopher born in 551 BC. His descendants can be traced back through 83 generations totalling more than two million people. In fact, it is quite easy to show that everyone alive today has a high probability of being genetically related to Confucius. Once we go back about a thousand years, we find that, statistically, we are related to everyone who was alive at that time. At a similar time period away in the future, everyone who is alive will be distantly related to your children and you!

Humans have a strong desire to be remembered after they die. We wish to leave something behind; we hope to live on in the memories of future generations. Memories of our grandparents and their parents live on in fragments, through word of mouth and some memorabilia. Once we die, knowledge of our past is quickly lost. Most of the events that take place in our lives will be forgotten within a few generations. The media on which we store things erodes, printed words fade, hard disks crash and corrode. If you want something to last a really long time, engrave it on tungsten, gold or platinum, depending on how much you can afford.

But why do we want to be remembered? Is it to give some meaning to our lives or is it just our human arrogance?

The legendary poet Sappho, considered by Plato to be the tenth muse, lived on the Greek island of Lesbos 2,600 years ago. Little is known about her life, but fragments of her writings remain: "And I say to you, someone will remember us, in time to come." We like to build monuments, a symbol of our achievements and a visual memory that stays in place long after our words and actions have been forgotten. The rare exceptions will be remembered for many centuries after their deaths for their contributions to science or art, literature or music. Others are remembered for their deeds to humanity, deeds that may be judged as good or evil according to our self-imposed rules of morality. Regardless, we have a desire for our actions, dreams and memories to live on.

Surely There Is More to All of This?

I am often struck by the overwhelming belief that exists in the minds of people that there is "more to all of this". Need there be more? Isn't the universe simply amazing enough? I hope I have shown how the scientific method that began with Thales, of reproducible experiment and logic, has led to our remarkable advancement. This is not just in our knowledge of how the universe works, but also in the technology that we take for granted in our lives. This does not stop the majority of the world's population from believing in things that have no scientific support and have even been systematically shown to be false. There is no basis and no evidence for astrology, telepathy, clairvoyance, reincarnation, ghosts, spirits, angels, witches, channelling, numerology, homeopathy, reflexology, alien abductions, paranormal activity, gods, demons, heaven or hell. The list of things that people want to believe in is long. Despite in-depth investigations and studies, blind tests and trials, there is no evidence to support any of them. As Mark Twain wrote in 1897, "Faith is believing what you know ain't so."

We are free to have our individual beliefs. They only become a problem when we believe in them so strongly that we try to impress them on others. I enjoy debating the existence of "more than there is" with my friends and family. But sometimes I feel I have to apologise because science kills these dreams. Unfortunately, many of these "beliefs" govern our lives to such an extent that we live our lives by them. Others may kill or die by them.

These reflections notwithstanding, let me give you some fuel for your dreams and imaginations. In the far, far future it is possible that our universe will recreate itself.

Dark energy is the energy associated with a given volume of space. If the vacuum of space has an energy level, it is possible that a lower energy state exists. This means that there is a possibility that a transition occurs from the energy state of space today into a lower energy state tomorrow. This could result in a new "inflationary phase" whereby patches of the universe expand exponentially, creating brand new universes. There is a possibility that this is what is happening to our visible universe right now. We cannot calculate the timescale over which this would occur since we do not yet have a theory that explains dark energy, but the possibility remains. Regions in empty space begin to transform to the new vacuum state. The fundamental parameters of our universe today, the masses of particles, the strength of the forces that attract and repel particles, the gravitational constant and the speed of light may all be different in these new island universes. Some may collapse instantly; others may expand forever. In most of these universes, the parameters are unlikely to be conducive to forming atoms, let alone stars and life. But in the possible infinity of space there may exist regions that are not too dissimilar from our own.

Our universe appears to be finely tuned to enable life to evolve and ultimately ask the questions we have. Many scientists have pointed out numerous cases in which small changes to the constants of nature would result in a universe inhospitable for life or even a universe without stars and planets. The "anthropic principle" is often evoked to make statistical arguments for why certain aspects of the universe are as they are. Many religious ideas and scientists' "strong" anthropic principle are based on the assumption that we are special and that the universe is just right for life as we know it. This reasoning is fallacious because it ignores the incredible capability of life to evolve and adapt to changing surroundings. It also ignores the possibility that life may have emerged elsewhere in our vast universe in a completely different form.

This concept of a multiverse removes the need for a designer who created our universe with the vision of it eventually leading to the existence of humans. There will be enough new universes appearing such that one will be just right for life! However, perhaps intelligent life can figure out a way to create designer universes that have the necessary conditions. In 1956 Isaac Asimov captured the essence of this notion in his profound short story "The Last Question". It begins: "The last question was asked for the first time, half in jest, on May 21, 2061, at a time when humanity first stepped into the light." In four pages Asimov takes us on a journey over billions of years into our future as mind and matter ultimately merge into one, a universal computer called Multivac. Its purpose is to answer the ultimate question of how to save humanity given the inevitable heat death of the universe that entropy delivers.

The Meaning of Life

Why are we here? Why is the universe here? What are we supposed to do with our short lives? What is the point of it all?

Who am I to question the meaning of life? But I am far from alone in asking this question. We have made a journey from the beginning of time to the end of the universe—our home and the home of all of our future descendants. If all life will ultimately die and nothing lasts forever, what indeed is the purpose?

I believe that there is no grand purpose. None.

It is not necessary to look to others or beyond for a purpose to life. Life has the purpose that you give to it. Our bodies are survival machines governed by our brains. The short-term aim is simply to preserve our genes, which may be altruistically motivated or not. Evolution did not follow a predesigned path that led to "intelligent thinking creatures". The first reproductive complex molecules were not aware of anything. They just occurred by chance, collections of atoms that obeyed the rules of physics—leading to selective processes that preserved certain beneficial traits. Until one day, one particular species of mammals happened to realise that they could use their brains to the full and appreciate that there are big questions to be asked and answered.

The emergence and proliferation of life on our planet is indeed an incredible thing. However, evolution has no purpose. It is just what it is: particles, atoms, molecules following the laws of physics. Life, like evolution, appears to be a random walk, despite the fact that predictability may underlie that apparent randomness. Decisions that seem random lead us to events that can change how we live the rest of our lives. We learn, we work, we love, we reproduce and we die. It is an incredible thing to be conscious and alive, even if it is for a brief moment in the infinite time ahead. It seems nothing less than a miracle.

Evolution follows no rules and laws except for the laws of physics. Evolution has no prescription for how to behave or how to be moral. Morality and rules are not something inherent in our genes—at birth our brains are a blank canvas sculpted by countless millions of years of evolution; that canvas is painted by our teachers and mentors, our role models and idols. Morality and rules are things we learn, rules that are invented to "make society better". Morality is not necessarily a consequence of evolution: Evolution does not say anything about how we should or should not use our brains. Our brains evolved for the purpose of survival, and only in the past 30,000 years have humans found a different use for them, such as creative thinking or pleasure seeking. Our natural evolutionary path has effectively been halted through our lifestyles, rules and morality.

There is nothing special about this time in which we live, 13.8 billion years after the beginning of our universe. There is nothing special about the star that provides our energy, or its position within the galaxy, or the place of our galaxy in the vast universe beyond. But it is rather incredible that on our little rock called Earth, life has evolved until it ultimately developed the skills to figure out everything I have recounted in these pages; life has managed to understand the basic laws of nature and a large part of the history and future of the universe.

The destiny of our lives, of humanity, is in our own hands. Make of it what you will. The destiny of life in the universe is constrained by the laws that govern our universe, and those laws will lead to the end of all life and the end of all consciousness. The relentless expansion of space, the steady goal of gravity to assemble and then disassemble, the conversion of matter into photons, the steady increase in entropy—all of this predicts a cold, empty, dark universe. Our distant future is one with no laughter, no love, no music, no warmth and no thoughts. Despite all of that, our universe and the life we see around us is indeed an amazing thing. All we can do is to live our short lives to the full, enjoy our brief moment of consciousness in this vast and spectacular universe. Make the most of your lives. Live your dreams.

Acknowledgements

This book came to life because of the inspiration, ideas, and support of others. I would like to thank my mother and father for everything and my lovely siblings, Janet, Billy, and Celia. Gabrielle, for reading and re-reading endless early versions of this text and helping me clarify all of those complicated parts, and for our wonderful children, Mariana and Joe. Suzanne Wilde for encouraging me to begin writing and for the creative suggestions for its contents. Katharina Blansjaar for the beautiful artwork, her intimate knowledge of language, and for all the next steps together in life. Joachim Stadel for all the creativity and fun and for helping turn my research dreams into a reality. Rolf Dobelli and ZurichMinds together with Peter Haag of Kein and Aber for making this happen. I would also like to thank all my colleagues and friends of the University of Zurich who contributed to this book in various ways, especially Oscar Agertz, Raymond Angélil, Natasha Arora, Aaron Boley, Jonathan Coles, Sebastian Elser, Tobias Goerdt, Lea Giordano, Nico Hamaus, Lawrence Krauss, George Lake, Aaron Manalaysay, Doug Potter, Justin Read, Prasenjit Saha, Romain Teyssier, and Marcel Zemp. And to all the people who have supported and inspired me during my random walk through life, thank you.

B. Moore, *Elephants in Space*, Astronomers' Universe,
DOI 10.1007/978-3-319-05672-2,

Glossary

Absolute zero Defined as zero Kelvin, the theoretical temperature at which entropy is at its minimum value. This is -273.15 degrees on the centigrade scale. In practice, it is impossible to achieve absolute zero. The coldest temperature ever achieved in a laboratory experiment is a billionth of one degree above absolute zero.

Algorithm A set of instructions that perform a calculation.

Antimatter Particles that have properties exactly opposite to normal matter. When a particle and its antiparticle collide, they annihilate each other and create new particles, usually photons.

Asteroid The solar system is filled with millions of asteroids that orbit the Sun. They are rocky objects larger than 10 metres across left over from the formation of the planets. The majority of asteroids orbit in the "asteroid belt" that lie between Jupiter and Mars.

Astrophysics The subject area that deals with the physics of the universe. It is closely related to astronomy, the study of the celestial objects beyond the Earth's atmosphere, and cosmology, the study of the origin and structure of the universe.

Atom The basic unit of matter from which stars, the Earth, and everything we see around us is made. The atom has a dense nucleus of protons and electrons that are bound together by the strong force, which is surrounded by a cloud of electrons that are bound to the nucleus through the electromagnetic force. The number of protons determines the chemical element, whereas the number of neutrons determines its isotope. If the number of electrons is the same as the number of protons, the atom has no net charge and is neutral. If these numbers differ, then it is called an ion.

Big bang The big bang model describes the origin and development of our universe, from a fraction of a second after time and space appeared, when it was incredibly hot and dense, through its subsequent expansion as it cooled and led to the formation of the stars and galaxies we observe around us today.

Black dwarf The ultimate state of a white dwarf that has cooled down to such an extent that it no longer emits significant radiation. It takes about 10^{15} years for a white dwarf to cool down to 5 Kelvin.

B. Moore, *Elephants in Space*, Astronomers' Universe,
DOI 10.1007/978-3-319-05672-2, © Springer International Publishing Switzerland 2014

Black hole When enough mass is concentrated in one place, it curves spacetime to such an extent that even light cannot escape or be reflected from its surface. In the eighteenth century, geologist John Michell speculated about the existence of black holes, although the fact that matter could influence the motion of light was shown by Einstein through his equations of general relativity.

Bosons Elementary particles are classified as either fermions or bosons. The five bosons that are observed to exist are photons (carriers of the electromagnetic force), the W and Z (carriers of the weak force), the Higgs boson and gluons (carriers of the strong force). Some models for quantum gravity predict the existence of a sixth boson, the graviton.

Brown dwarf Brown dwarfs are failed stars which were not massive enough to generate sufficient temperatures during their gravitational collapse to ignite hydrogen fusion. The critical mass for this to occur is about 8 percent of the mass of the Sun.

CERN: European Organization for Nuclear Research An international scientific collaboration based in Geneva, Switzerland, that hosts particle accelerators used for carrying out research in high-energy physics. The World Wide Web was founded in 1989 at CERN by Tim Berners-Lee. The main experiment today is the Large Hadron Collider.

Chaos A chaotic system behaves in an unpredictable way and is very sensitive to small changes in the parameters that determine its behaviour.

Classical (e.g. classical physics) Classical physics is based on Newtonian mechanics and predates relativity and quantum mechanics.

CMB: Cosmic microwave background Photons produced 380,000 years after the big bang move freely in all directions, giving rise today to a background of low-energy light that appears to come from all directions. It contains a wealth of information about the early universe as well as the development of gravitational structure.

Comet Comets are icy objects that form and orbit at large distances from the Sun, but sometimes their orbits are perturbed and they come into the inner solar system. If they come close enough to the Sun, debris tails can be seen sweeping behind them caused by solar wind and solar radiation evaporating material from their surface.

Constants of nature Parameters that set the absolute values of certain quantities, such as force and charge, or fundamental constants, such as the speed of light. It is not known what sets these parameters, and they seem to be finely tuned to enable structure to form in our universe.

Conservation laws Quantities associated with an isolated physical system that do not change with time.

Cosmological constant A parameter originally introduced by Einstein that provides a negative pressure enabling a static and stable universe. The same parameter may describe the observed accelerated expansion of the universe and may be a property of space itself.

Cosmology The study of the universe as a whole, its origin, and evolution.

CPU The central processing unit of a computer that carries out mathematical calculations and interfaces with other components such as disk storage and graphics cards.

Dark energy The energy associated with a given volume of space that is hypothesised to exist to explain the observed accelerated expansion of the universe.

Dark matter Mass that is identified as existing within and around galaxies and galaxy clusters but whose nature is unknown. It is called dark matter because it is not made of baryonic particles that would emit photons.

Density The amount of mass within a given volume, e.g. kilograms per cubic metre.

DNA Deoxyribonucleic acid contains the genetic instructions for the development and functioning of all known life forms. It is made of long strands of molecules arranged in a double helix and in humans it is several billion base pairs (rungs of the helix ladder) long.

Electromagnetic force This is one of the four fundamental forces of nature. It acts between charged particles and gives rise to a host of phenomenon such as electricity, magnetism, and chemistry.

Electromagnetic spectrum Light, or photons, can have a range of wavelengths (and corresponding frequencies), and the electromagnetic spectrum is a way of classifying photons according to their wavelengths.

Electron An electron is an elementary particle that has a negative charge exactly opposite to that of a proton and a mass that is 9.1×10^{-31} kilograms, which is 1/1836 that of a proton.

Equilibrium In physics an equilibrium is achieved when all the forces acting on a system are balanced and the global state of the system does not change with time.

Fermions Particles such as electrons or protons that are usually associated with matter are called fermions. They have a property called spin that is equal to half an integer and are subdivided into two further classes, leptons, and quarks. The electron is an elementary lepton particle, whereas protons and neutrons are composite particles made of quarks.

Fission The process by which the nucleus of an atom splits into lighter nuclei and a fraction of the initial rest mass energy is released in the form of high-energy photons.

Frequency The number of events within a given interval of time. A photon can be considered as an oscillating electric and magnetic field that travels through space and the frequency describes the number of oscillations per second. The wavelength of light is inversely proportional to its frequency.

Fusion The process by which atomic nuclei coalesce and become one heavier nucleus which releases energy, mainly in the form of neutrinos and high-energy photons.

Inertia The resistance of any physical object to a change in its motion which is proportional to its mass.

Galaxy A collection of stars and gas embedded within an extended distribution of dark matter.

General relativity A description of gravity that incorporates the properties of space and time which generalises the separate descriptions of special relativity and Newtonian gravity.

Gene A section of DNA that contains information about how to build and maintain cells and how to pass on traits to offspring. The gene is a unit of inheritance.

G, gravitational force constant Matter attracts other matter with a force that is proportional to the mass. The gravitational force constant tells us just how much force a given amount of matter exerts.

Gravitational lensing The gravitational field of massive objects distorts space and time, and photons follow curved paths around them. Background objects can be magnified and distorted in a way similar to how a glass lens deflects light.

Greenhouse effect An increase in the surface temperature of the Earth due to molecules in the atmosphere that trap the re-radiated heat.

Habitable zone The orbital zone around a star where a planet could hold liquid water on its surface without it boiling away or freezing.

Hadron Hadrons are composite particles, such as neutrons and protons, that are made of quarks held together by the strong force.

Homogeneous When something is the same in all places.

Homo sapiens Anatomically modern humans who appeared in Africa less than two hundred thousand years ago.

Hubble's law The observation that galaxies are receding from our own at a rate that is proportional to their distance away.

Inflation A rapid exponential increase in the size of the universe during the first 10^{-32} seconds of its existence that is hypothesised to occur to explain certain observable features of our universe.

Interstellar medium The diffuse gaseous medium that exists in galaxies and fills most of the space within them. It provides the material from which new stars form.

Isotope A variant of an atom that contains a different number of neutrons in its nucleus.

Isotropic When something is the same in all directions.

Kelvin A unit by which scientists measure temperature which is one of the internationally accepted standard (SI) units.

Kinetic energy The energy that an object has due to its motion relative to something else.

Logic gate A device that performs a logical operation on one or more inputs which produces a single output.

Luminosity The amount of energy that an object radiates per unit interval of time.

Meteor Rocky objects in the solar system that are smaller than about 10 metres across are called meteoroids. When they enter the Earth's atmosphere, they are vapourised and leave a brief visible trace in the sky, becoming meteors (shooting stars).

Meteoroid A small space rock less than 10 metres across.

Meteorite The debris of a meteor which reaches the Earth's surface.

Momentum A property of moving objects that is equal to the product of their mass times their velocity. The rate of change of momentum is equal to the force that is causing the change.

Neanderthal An extinct member of the *Homo* genus that existed from several hundred thousand years ago to as recently as 30,000 BC.

Nucleosynthesis The process by which the nuclei of hydrogen, helium, and lithium formed during the first minutes following the big bang.

Neuron A complex cell that processes and transmits information through electrical and chemical signalling.

Neutron A subatomic hadron particle made of three quarks that has no net charge and a mass of 1.675×10^{-27} kilograms, which is about one percent larger than that of the proton.

Neutron star The dense stellar remnant that results from the gravitational collapse of a massive star which compresses the remaining material to the density of atomic nuclei.

Neutrino An elementary fermion particle that is neutral has very low mass and travels close to the speed of light. They are typically produced in nuclear reactions and radioactive decays.

Newton's inverse square law (Newtonian gravity) The force of gravity between two objects is proportional to the product of the two masses divided by the square of the distance between them. The constant of proportionality is the gravitational constant, G.

Newton's laws of motion (i) A body remains at rest or in uniform motion at a constant velocity unless acted on by a force. (ii) The rate of change of momentum of a body equals the force acting upon it. (iii) If two bodies exert a force on each other, the forces are equal in magnitude and opposite in direction.

Opposable thumbs Animals which have thumbs that can face their fingers and grasp things. This is common to many primates but also other animals such as pandas and possums. Some species also have opposable toes.

Palaeolithic The era during which our ancestors used stone tools that stretches from three million years ago through to the Mesolithic era that followed the last great Ice Age in 10,000 BC.

Paradox A self-contradictory assertion or situation resulting from a seemingly true statement which seems to defy logic.

Photon A quantum of radiation that transmits the electromagnetic force and which could be considered as a particle with zero rest mass and charge. The energy of a photon equals Planck's constant multiplied by its frequency.

Planck's constant A physical constant of nature which is related to the energy in a single quantum of radiation.

Pulsar A rapidly rotating neutron star that emits radiation mostly in the form of radio waves which are powered by an intense magnetic field. The magnetic field is misaligned with the rotational axis of the neutron star which makes the beams of light sweep around like the beacon of a lighthouse. The interval between the pulses that we observe is as regular as an atomic clock.

Proto-planetary disk A rotating disk of gas and dust that surrounds a newly forming star. The gas contains most of the atoms in the periodic table, and it is within this medium that the planets form.

Proto-star A star that is just forming through gravitational collapse that glows brightly because of the energy released in the collapse, but is not yet hot enough to allow nuclear fusion to occur.

Proton A subatomic hadron particle composed of three quarks that is one of the fundamental constituents of atomic nuclei. It has a mass of 1.673×10^{-27} kilograms and a positive charge that is exactly opposite in strength to that of the electron.

Quantum mechanics A branch of physics that deals with the nature of light and particles and describes them mathematically with a probabilistic wave equation.

Quarks An elementary particle that combines in triplets to form hadron particles, such as protons and neutrons.

Red dwarf A star with a mass less than about half that of the Sun and extending all the way to the hydrogen-burning limit of about 10 percent of the Sun's mass. Red dwarf stars evolve very slowly and live for a trillion years, but they never become hot enough to begin the fusion of helium.

Space I cannot define space!

Singularity A point in space which has infinite density resulting from matter which occupies an infinitesimally small volume. It is a mathematical statement that results from our lack of understanding of how matter behaves once it reaches extreme densities.

Special relativity A theory that connects together space and time that was developed by Einstein based on the assumption that the speed of light is a constant in all reference frames.

Spectrum The intensity of light as a function of its frequency or wavelength. Our eyes are sensitive to the optical part of the electromagnetic spectrum, which we perceive as a colour.

Star A massive gravity-bound plasma of atoms that is hot and dense enough to sustain fusion reactions between some of its elements.

Strong force Also called the strong interaction, it is one of the four fundamental forces of nature. It binds together neutrons and protons allowing the nuclei of atoms to form.

Supernova The explosion that occurs during the last few seconds of the evolution of massive stars as their iron cores collapse at close to the speed of light. The energy produced in the collapse is similar to the energy produced over the entire lifetime of the stars. Consequently, they can be seen across most of the visible universe.

Symmetry Symmetries in geometry and physics describe how an object appears similar upon a coordinate change, such as reflection (mirror image) or rotation.

Turbulence The complex flow behaviour of gases and liquids under certain conditions.

Universe Everything that exists: all of space, time, and matter. The visible universe is defined as the region light has had time to travel across over its 13.8 billion year history.

Wavelength Light can be considered as an oscillating electric and magnetic field that travels through space. The distance between successive peaks of the wave is known as the wavelength. The wavelength of light is inversely proportional to its frequency.

Weak force Also called the weak interaction, it is one of the four fundamental forces of nature. It is important in the fusion processes in stars and in the radioactive decay of particles.

WIMPs Weakly interacting massive particles. Candidates for dark matter that interact very weakly with atomic matter but have yet to be detected in laboratory experiments.

White dwarf When most stars end their lives and stop nuclear fusion, their remaining matter condenses into a dense state and they radiate away their remaining energy. Eventually, they become cold, invisible black dwarfs.

Further Reading

1. Adams FC, Laughlin G (2000) The five ages of the universe: inside the physics of eternity. Free Press, New York
2. Aristotle (2008) In: Bostock D (ed) Physics (trans Robin Waterfield). Oxford University Press, Oxford
3. Asimov I (1994) The last question. In: Complete stories, vol 1. HarperCollins, London
4. Copernicus N (1995) On the revolutions of heavenly spheres. Prometheus Books, Amherst, NY
5. Darwin C (2012) On the origin of species by means of natural selection or the preservation of favoured races in the struggle for life, 2nd edn. Tredition Classics
6. Richard D (2006) The selfish gene, 30th Anniversary edn. Oxford University Press, Oxford
7. Richard D (2009) The greatest show on Earth: the evidence for evolution. Bantam Books, London
8. Feynman R, Leighton RB, Sands M (2011) The Feynman lectures on physics, The New Millennium edn. Perseus Distribution, Jackson, TN
9. Krauss L (2012) A Universe from nothing. Simon & Schuster, New York
10. Newton SI, Cohen BI, Whitman A, Budenz J (1999) The Principia. University of California Press, Oakland, CA
11. Russell B (2011) A free man's worship. In: Russell B, Burroughs J (eds) Two modern essays on religion. Literary Licensing, Whitefish, MT
12. Weinberg S (1993) The first three minutes: a modern view of the origin of the universe. Basic Books, New York

B. Moore, *Elephants in Space*, Astronomers' Universe,
DOI 10.1007/978-3-319-05672-2,

Printed by Publishers' Graphics LLC